SAILING
FROM POLIS TO EMPIRE

Sailing from Polis to Empire

Ships in the Eastern Mediterranean During the Hellenistic Period

Edited by Emmanuel Nantet

https://www.openbookpublishers.com

© 2020 Emmanuel Nantet. Copyright of individual chapters is maintained by the chapters' authors.

This work is licensed under a Creative Commons Attribution 4.0 International license (CC BY 4.0). This license allows you to share, copy, distribute and transmit the text; to adapt the text and to make commercial use of the text providing attribution is made to the authors (but not in any way that suggests that they endorse you or your use of the work). Attribution should include the following information:

Emmanuel Nantet (ed.), *Sailing from Polis to Empire: Ships in the Eastern Mediterranean During the Hellenistic Period*. Cambridge, UK: Open Book Publishers, 2020, https://doi.org/10.11647/OBP.0167

In order to access detailed and updated information on the license, please visit https://doi.org/10.11647/OBP.0167#copyright

Further details about CC BY licenses are available at https://creativecommons.org/licenses/by/4.0/

All external links were active at the time of publication unless otherwise stated and have been archived via the Internet Archive Wayback Machine at https://archive.org/web

Any digital material and resources associated with this volume can be found at https://doi.org/10.11647/OBP.0167#resources

Every effort has been made to identify and contact copyright holders and any omission or error will be corrected if notification is made to the publisher.

ISBN Paperback: 978-1-78374-693-4
ISBN Hardback: 978-1-78374-694-1
ISBN Digital (PDF): 978-1-78374-695-8
ISBN Digital ebook (epub): 978-1-78374-696-5
ISBN Digital ebook (mobi): 978-1-78374-697-2
ISBN XML: 978-1-78374-698-9
DOI: 10.11647/OBP.0167

Cover image: Delos, House of Dionysos, Room L, Eastern Wall (1st century BCE): graffito of an Hellenistic warship with 85 oars (drawing by Dominique Carlini in *Récit d'une aventure : les graffiti marins de Délos : Musée d'histoire de Marseille, 18 décembre 1992 – 22 mars 1993*, Marseilles, Marseilles Historical Museum, 1992). All rights reserved.

Cover design: Anna Gatti.

Contents

Preliminary Notes	vii
Authors	ix
Preface	xi
Alain Bresson	
Bibliography	xvii

1. The Hellenistic Merchantmen: A Contribution to the Study of the Mediterranean Economies — 1
   Emmanuel Nantet
   Bibliography — 6

2. Evolutions of the Representation of the Eastern Mediterranean in the Hellenistic Period — 11
   Jean-Marie Kowalski
   2.1. Granularity of Information — 14
   2.2. Distances and Maritime Experience — 16
   2.3. Seasonality of Weather Indications — 18
   2.4. Influence of Weather Conditions Over Navigation — 20
   2.5. Granularity and Quality of Information: The Problem of Salience — 21
   2.6. Salience and Visually Distinctive Features: The Case of Cape Pedalion — 24
   2.7. Conclusion — 24
   Bibliography — 25

3. Naval Architecture. The Hellenistic Hull Design: Origin and Evolution — 27
   Patrice Pomey
   Bibliography — 51

4.	*Naves Pingere*: 'Painting Ships' in the Hellenistic Period *Martin Galinier and Emmanuel Nantet*	55
	4.1 Naval Issues Before the Reign of Alexander	56
	4.2. Ship Painters	60
	4.3. Conclusion	67
	Bibliography	68
5.	The Rise of the Tonnage in the Hellenistic Period *Emmanuel Nantet*	75
	5.1. The Sources	76
	5.2. An Initial Rise in the First Part of the Second Century?	80
	5.3. A Second Rise from the End of the Second Century to the Beginning of the First Century?	82
	5.4. The Common Reasons for the Two Increases	84
	5.5. Conclusion	86
	Bibliography	86
6.	A Note on the Navigation Space of the *Baris*-Type Ships from Thonis-Heracleion *Alexander Belov*	91
	6.1. Main Characteristics of the *Baris* as per Herodotus and New Archaeological Data	93
	6.2. Navigation Area of the *Baris*-Type Ships	94
	6.3. Conclusions	106
	Bibliography	110

List of Tables and Illustrations	119
Index	123

Preliminary Notes

This book originated partly from an international workshop, which was held on 14 November 2014 in Nicosia, at the Archaeological Research Unit of the University of Cyprus. It was organized by Stella Demesticha (University of Cyprus) and Emmanuel Nantet (then University of Le Mans, France, now at the University of Haifa, Israel, based in The Leon Recanati Institute for Maritime Studies, Laboratory of Nautical Archaeology and History). It was supported by the 'Sailing in Cyprus Through the Centuries' project, the French Institute in Cyprus (Institut Français de Chypre), the CReAAH (Centre de Recherches en Archéologie, Archéosciences, Histoire — UMR 6566) and the Scientific Interest Group of Maritime History (Groupement d'Intérêt Scientifique d'Histoire Maritime).

The book has been edited by Emmanuel Nantet and improved by further contributions, with sincere thanks to the contribution of the French Institute in Cyprus, of the CRESEM (Centre de Recherche sur les Sociétés et les Environnements Méditerranéens — EA 7397) and the LabEx Archimède (Archéologie et Histoire de la Méditerranée et de l'Égypte Anciennes). This book has benefited greatly from the advice of Conor Trainor (University of Warwick) and Christoph Schäfer (University of Trier). It has been copy-edited and proof-read by Sharon Elisheva Turkington, Ivana Kubalova and Lucy Barnes.

Authors

Alexander Belov, Centre for Egyptological Studies of the Russian Academy of Sciences, Moscow.

Alain Bresson, University of Chicago.

Martin Galinier, Université de Perpignan Via Domitia, CRESEM E.A. 7397.

Jean-Marie Kowalski, University of Paris-Sorbonne / French Naval Academy FED 4124.

Emmanuel Nantet, Department of Maritime Civilizations, Laboratory for Nautical Archaeology and History, The Leon Recanati Institute for Maritime Studies, University of Haifa, associated member UMR 6566 (CReAAH).

Patrice Pomey, Emeritus Research Director, AMU, CNRS, MCC, Centre Camille Jullian.

Preface
Alain Bresson

The absence of technological progress in the ancient world has long been a dogmatic belief among ancient historians, linked to the idea that the ancient economy was stagnant. It took time, and also a prolonged and vigorous debate, to explode both pronouncements. Recent research has shown that starting in the Archaic period, and culminating at the end of the Hellenistic period and at the very beginning of the Imperial period, the ancient Mediterranean world experienced a vigorous period of growth. The evidence for this process is abundant and manifold: from the basic quantity of ceramic shards on archaeological sites to the size of houses and cities, or the number of various artefacts found in these sites.

Admittedly, the idea has also long prevailed that, to the extent that there was growth, it was purely the consequence of demographic expansion rather than the result of any productivity increase. But this idea also must be abandoned. Economic growth in the ancient world was fundamentally based on a specific institutional organization, that of the city, which firmly guaranteed property rights. This meant property rights over land and any other material item, but also over people, slavery being one of the pillars of ancient society. Some would even (wrongly) argue that the exploitation of enslaved men and women was the only fuel of economic growth. But no matter what, if an analysis of the factors of economic growth must include the diverse forms of exploitation of the workforce, it should not neglect technological progress and innovation. Indeed, the process of growth was also based on a comparatively vigorous technological progress. The fact that the ancient world did not introduce the steam engine (and other technologies that harness huge quantities of energy) has seemed to condemn all the technological progress that took place during this period. Technological innovation in the ancient world was less spectacular than that of the modern period, as the latter is the result of a systematic combination of scientific knowledge and technological developments. Yet, in various sectors of the ancient economy, the process of innovation achieved impressive results, which allow us to understand how economic growth could actually take place. One of the major technological breakthroughs of the ancient world occurred in sailing technology. In this respect, both for its quantitative and qualitative aspects, naval archaeology provides a major contribution to our understanding of this phenomenon.

For the former, one can think of the now famous graph produced by Anthony J. Parker, which, since its introduction into the scholarly debate, has been regularly updated without radically changing the overall picture. The graph of the number of shipwrecks between the Archaic period and Late Antiquity has a Gaussian aspect. One might argue that the graph illustrates the growth of trade relations, not global economic growth per se. One could also contend that real economic growth did not follow such an abrupt pattern of increase and decline. This is not the place to address these questions. Nonetheless, given that the growth of the ancient economy was directly linked to the expansion of trade, primarily maritime trade, the graph of shipwrecks illustrates the process of economic growth (and decline from the second century CE onwards).

As for maritime trade, recent research has shown that the technology of shipbuilding experienced several major transformations during Classical antiquity. That is where this volume, *Sailing from Polis to Empire: Ships in the Eastern Mediterranean during the Hellenistic Period*, is important. It publishes the papers presented at an international workshop that took place in Nicosia, at the Archaeological Research Unit of the University of Cyprus, on 14 November 2014. This workshop was organized by Stella Demesticha (University of Cyprus) and Emmanuel Nantet (then Université du Mans, now University of Haifa). Emmanuel Nantet recently published his monumental and justifiably acclaimed *Phortia: le tonnage des navires de commerce en Méditerranée du VIIe siècle av. l'ère chrétienne au VIIe siècle de l'ère chrétienne* (Rennes 2016: Presses Universitaires de Rennes), which was devoted to the question of ship tonnage in the ancient world. There is a clear complementarity between the two books. Of course, the diversity in authorship in an edited volume like this one also means a diversity of approaches to ancient naval archaeology. But the common thread is the ship, the 'forgotten hero' of the study of ancient economic life, as is emphasized from the start by Emmanuel Nantet himself.

    The chapter by Patrice Pomey, one of the scholars who has contributed most to the study of this technology, perfectly summarizes the various phases in the ancient technology of shipbuilding. The basic technology used for assembling ships in the Archaic Greek world was that of stitching. At the turn of the Archaic and Classical periods, the 'sewn ships' were replaced by ships assembled by tenon and mortise. This technique originated from Phoenicia and migrated westward. The Mediterranean world was not only an area where the accelerated transfer of goods could occur. It also provided ideal conditions for the migration of technologies, and unsurprisingly, given the direct link provided by the movement of ships and sailors, the technology of shipbuilding was one that could most easily migrate. With its tripartite structure—keel, planking, framing—the ship of the 'Hellenistic type' (as it is defined by Pomey) was still of the 'shell-first' variety. It was however much sturdier than its predecessor. Its size and its hollow shape (defined as a 'wineglass profile') meant that its tonnage could easily reach several hundred, as compared to the less than thirty of the early Archaic sewn ships. The small ships of the early Archaic period were fit for transporting mainly small quantities of luxury goods for wealthy elites, whereas the massive increase in the tonnage of ships made it possible to achieve the pan-Mediterranean long-distance transport of heavy freight loads for ordinary customers.

    The ship of the 'Hellenistic type' still had weaknesses. For instance, the keel was not firmly linked to the other parts of the structure and it could easily be lost after a shock, precipitating the inevitable sinking of the ship. Ships of the Imperial period, with their keelsons and several lateral sister-keelsons, were apparently more robust. Pomey's argument is supported throughout by a large number of illustrations (photos and drawings) and the reader can easily follow the demonstration. One can only be struck that the observations made on the shipwrecks match the ships depicted in Roman representational art so well, which in return helps the archaeologists reconstruct the often-missing parts of the wrecked ships, such as the prow or the upper parts of the hull. Emmanuel Nantet himself sees two main phases in the process of the growth of the tonnage of the Hellenistic ships: the beginning of the second century BCE, where this growth was pan-Mediterranean, and the turn of the first century BCE, where it was limited to specific routes and specific products like wine or works of art, directly connected with the new phase of the Roman conquest. Beyond the technological change in ship building, he also insists on the structural transformations in harbour construction necessitated by the increase in the size of merchantmen fleets and in the tonnage of their respective ships. This is currently one of the most active fronts of research in ancient navigation and nautical archaeology, as is made clear by the many and ground-breaking studies of Pascal Arnaud, on the institutional and practical side of access to ports, and Simon Keay, on port archaeology and specifically on Portus Romae, the imperial Roman port built in the first century CE.

Another side of ancient water transport is river navigation. Alexander Belov revisits the case of the *baris*. This type of ship is mentioned by Herodotus (2.179) when he explains that it was used in the internal waterways of Egypt. The word *baris* comes from the ancient Egyptian *br* (*byr*) and during the Eighteenth Dynasty it was a sea-going ship. But later, in Herodotus's time and until the Late Hellenistic period (the last mention in papyri is from 125 BCE), this ship was the typical Nile freighter. The case of the *baris* is fascinating because the textual evidence can be combined with excavation data. Indeed, the site of Thonis-Heracleion, at the mouth of the Canopic branch of the Nile, has proved to be a gold mine for our understanding of ancient navigation and shipbuilding. The site has been explored by Frank Goddio and his team for the last two decades. The underwater excavation has revealed a large number of shipwrecks. Belov himself participated in the exploration of the site and has a first-hand knowledge of the material. More than sixty ships have been definitively identified but their actual number is certainly significantly higher. Some of these shipwrecks, like Ship 17, allow us to form a vivid picture of these craft.

Belov has devoted a monograph to this ship. It was built of local wood (acacia) and had no proper keel, which was not a problem for Nile navigation but rather an advantage. Such a ship had to be hauled upstream. It was 27–28 m long and its tonnage was c. 113 metric tons.

Let us stress that this was a considerable amount for this period. If we apply the rule that one *medimnos* of wheat (the standard grain production of Egypt) weighed 31 kg, the cargo of a *baris* was equivalent to 3645 *medimnoi*. This was slightly over the capacity of a standard Greek sea-going ship of the Classical period (c. 3000 *medimnoi*). The cargo of a single Nilotic *baris* would have easily filled the hold of a Greek sea-going ship bound for its homeland. Once again, we see how important it is for the historian to combine textual evidence and archaeological data.

Navigation and sea routes are also considered in this volume. Jean-Marie Kowalski analyses the navigation routes from and to Cyprus on the basis of literary sources, from Herodotus through to Strabo and the *Stadiasmus maris Magni*. It is important, as Kowalski does, to use the data provided by our literary sources not *in abstracto* but in their geographic and ecological framework. This implies taking into account the differences in the wind directions between the summer and winter seasons. From this perspective, it is perfectly legitimate to use modern climatic data to make sense of the ancient literary sources, as it has been done for the conditions of navigation around the Triopion (cape Krio, Knidos).

Another aspect of the life of ships — their decoration — is addressed in Martin Galinier and Emmanuel Nantet's chapter. 'Painting vessels' could have two meanings in ancient tradition: depicting vessels in a painting or actually painting vessels. Building on an anecdote related by Pliny (*NH* 35.101) about the life of the famous painter Protogenes of Kaunos, Galinier and Nantet cleverly offer a small masterpiece, an analysis in the form of a diptych covering both aspects of 'painting vessels.' The depiction of vessels in the ancient pictorial tradition was illustrated by vase painters and also by the most famous masters like Apelles and Protogenes. Pliny informs us that, to earn his living, Protogenes began his career as a vessel painter. Many texts, as well as pieces of representational art such as paintings and coins, confirm that ancient ships were lavishly decorated, and for this reason there should be no doubt about the actual meaning of Pliny's allusion: before representing ships in his paintings, Protogenes had been a simple ship painter. Indeed, the ships were adorned with reliefs painted in bright colours. The painting often consisted of tinted wax, with additives allowing the mix to resist the effects of sun and salt water. Ruddle or red ochre (*miltos*), a dye that was supposed to protect the wood from decay caused by worms, was used to paint ancient ships. It was for this reason that, in the fourth century BCE, Athens established its monopoly over the island of Keos in the Cyclades, which was a large producer of this pigment.

Beyond protecting the wood, it remains clear that the decoration of ships, especially of warships, was seen as a standard part of their equipment. This is true not only for antiquity but also for the Western tradition at least until the end of the early modern period. Until war became an industrial process in the course of the nineteenth century, going to war both on land and sea was also a form of pageantry. For ships, this meant displaying a spectacular array of colours and reliefs in order to capture the imagination of both seamen and landsmen, friends or foes. The most stunning testimony of this practice remains the Swedish warship *Vasa*, shipwrecked on 10 August 1628 during her maiden voyage, after navigating less than one mile from her port in the bay of Stockholm. The shipwreck was located in the 1950s and salvaged in 1961, and the *Vasa* is now on display at the Vasa Museum in Stockholm. Visitors can discover the hull and rigging, but they can also behold the many statues that decorated the ship, especially on the prow and stern portions. A careful examination of the ship's wood has resulted in the recovery of traces of pigments, allowing researchers to propose restorations of the original paintings. Visitors can thus admire on a replica the vivid colours applied to the decorations of the ship, allowing them to get a fair idea of the taste for the spectacular that at the time went with building a man-of-war.

As observed by Galinier and Nantet, who usefully quote Euripides' *Iphigenia in Aulis* (231–276), the decorations of the ships appealed to the imagination of the observers and a fleet parade was a show in itself. One understands even better the spectacle offered by the Athenian fleet leaving for Syracuse in June 415 BCE, as described by Thucydides (6.31.1–6), who emphasizes the expensive figureheads (*sēmeia*) and equipment of the vessels (6.31.3).

Obviously, this volume is important for economic historians, but also for scholars of social and cultural history. If nautical history has been long dominated by specialists of the Western Mediterranean, the balance is currently changing, as proved by this publication. The editor and contributors of this volume must be praised for that and encouraged to undertake further research in the same direction.

Bibliography

Bresson, A. 2011. 'Naviguer au large du cap Triopion.' *Anatolia Antiqua* 19:395–409.

Bresson, A. 2016. *The Making of the Ancient Greek Economy: Institutions, Markets, and Growth in the City-States*. Princeton: Princeton University Press.

Carrara, A. 2014. 'À la poursuite de l'ocre kéienne (*IG* II^2 1128): mesures économiques et formes de domination athénienne dans les Cyclades au IVe s. a.C.' In *Pouvoirs, îles et mers: formes et modalités de l'hégémonie dans les Cyclades antiques (VIIe s. a.C. — IIIe s. p.C)*, edited by G. Bonnin and E. Le Quéré, 295–316. Bordeaux: Ausonius.

Lytle, E. 2013. 'From Farmers into Sailors: Ship Maintenance, Greek Agriculture, and the Athenian Monopoly on Kean Ruddle (*IG* I^2 1128).' *Greek, Roman and Byzantine Studies* 53:520–550.

Parker, A. J. 1992. *Ancient Shipwrecks of the Mediterranean & the Roman Provinces*. BAR International Series 580. Oxford: Tempus Reparatum.

1. The Hellenistic Merchantmen

A Contribution to the Study of the Mediterranean Economies

Emmanuel Nantet

> Although numerous scholars have explored the Mediterranean economy of the last centuries BCE, their research has included hardly any data about shipwrecks. This can be explained not only by the lack of such data, but likewise the lack of conferences dedicated to this issue. The impact of shipwrecks on the Hellenistic sea trade is therefore a gap in our collective knowledge. The purpose of this book is to suggest some approaches to the study of this issue.

Since Rostovtzeff,[1] many scholars have shown an interest in the sea trade during the Hellenistic period. But like Finley,[2] most of the economic analysis of the Greek world deals primarily with the Archaic and the Classical periods.[3] The economy of the Hellenistic period suffers from a lack of rigorous analysis. Fortunately, some studies have been dedicated to the Hellenistic economies, but almost all of them focus on a kingdom[4] or a city. Of course, the royal power and the *polis* constitute the principal framework in which economies were strongly embedded. Nevertheless, this regional approach tends to overlook the

---

1 Rostovtzeff 1941, 2:1248–71.
2 Finley 1985.
3 See recently, Harris et al. 2016, who focus on the Classical period.
4 Préaux 1939; Chankowski and Duyrat 2004.

Mediterranean scale.⁵ This is why a group of scholars, including Zosia H. Archibald, John K. Davies and Vincent Gabrielsen,⁶ undertook to organize a series of three conferences to explore the Hellenistic economies.⁷ These conferences produced many case studies about the various issues at hand. The work does not rely on regional areas, which would limit the discussion, but on a thematic and comparative approach. Moreover, they rely both on written and archaeological evidence. These conferences have produced many fruitful works, which have improved our knowledge of this issue.

However, among the numerous works about Hellenistic economies, very few mention the ships.⁸ Only the conference on Hellenistic Economies at Liverpool in 1998 dedicated a paper to this issue⁹ and so far, Gibbins' article seems to be the only study focusing on Hellenistic shipwrecks. His paper is well documented and offers a useful appendix that consists of a list of sixty-four Hellenistic 'shipwrecks'¹⁰ relying on the data gathered by Anthony Parker. Although Gibbins' study is entitled 'Hellenistic Shipwrecks', it focuses only on the amphorae that the ships were carrying. Almost nothing is said about the hulls, apart from a few details about the hull of Kyrenia,¹¹ and a brief mention of the hull of Apollonia (discovered off the coast of Libya) in the appendix.¹² Furthermore, Gibbins does not write a word about the Ma'agan Mikhael shipwreck despite the fact that he deals with Classical shipwrecks — his article is very focused on the Aegean and Cypriot areas.

The second and third conferences about the Hellenistic economies did not include any contributions about ships either; despite this, they were mentioned from time to time,¹³ which shows how significant they are for our understanding of the sea trade. All the other contributions

---

5   For example, Scheidel et al. 2008.
6   And Graham J. Oliver at the first conference.
7   Archibald et al. 2001, 2005, 2011.
8   Note a well-documented exception: Bresson 2016, 86–88.
9   David Gibbins, 'Shipwrecks and Hellenistic Trade,' in Zofia H. Archibald et al. (eds.), *Hellenistic Economies* (London/New York: Routledge, 2001), pp. 273–312.
10 Ibid. 296–304, table 10.A1.
11 Ibid., 288–89.
12 Ibid., 297 (n°7).
13 For example, Bresson 2011, who often discusses shipping issues. Two recent chapters by this author provide interesting contributions to the study of maritime trade in the Hellenistic period: Bresson 2018a and 2018b.

to these conferences that discussed underwater remains dealt only with transport amphorae — for example, when Mark Lawall mentioned shipwrecks, he focused on amphorae cargoes only.[14] Even though these studies about transport amphorae are quite useful, the ship is systematically omitted: the resulting publications do not neglect data collected from underwater archaeology, but they deal mainly with cargoes, not hulls.

The ship, as a vital tool for commerce, is therefore missing from the global analysis, despite the fact that the study of the ship could contribute a great deal to our understanding of Hellenistic maritime commerce. Indeed, it allows us to measure the scale of trade. This must be understood in its geographic context.[15] Firstly, the sea trade relies on three kinds of maritime routes. The regional one, which joins Ephesus to Piraeus for example, or Alexandria to Rhodes, is well documented by various sources. However, our knowledge about the inter-regional route, which connected distant harbours of the Mediterranean such as Pozzuoli and Alexandria, relies almost exclusively upon written evidence. As for local routes, which linked harbours that lay only a few nautical miles apart, it is very hard to identify these. Nonetheless, the importance of short journeys must not be overlooked. In addition to these varying geographical scales, we also need to take into account the quantitative ones.[16] Were these amphorae embedded in lively or less active networks of trade? It is tempting to suggest that the bigger ships were carrying merchandise on long-distance trade routes between large and significant harbours, and that smaller ships were just redistributing the goods into the secondary harbours. This situation was certainly common, but it was not always so.

Thus, this inquiry about the varying nature of the sea trade raises many questions: how were these ships built? How big were they? How much could they carry? What merchandise did they convey? What was their navigation area? Where did these ships sail to and from? Was the situation different in the Eastern and the Western waters? A close examination of these ships will therefore contribute greatly to our understanding of Hellenistic economies.

---

14 Lawall 2005, 191.
15 Nantet 2016, 171–73.
16 Ibid.,173–75.

So far, most scholars who have dealt with Hellenistic ships have not focused closely on economic issues.[17] Since there were almost no shipwreck remains in the Eastern Mediterranean, there was very limited information about cargoes; only iconography and written evidence was available, so architectural features and legal issues have been to the fore. Whereas research on the Western Mediterranean put technical and economic issues at the heart of maritime studies,[18] those focusing on the Eastern region have answered different questions. This lack of interest in sea trade among the scholars who have studied Hellenistic ships in the Eastern Mediterranean was also not conducive to the analysis of Hellenistic economies.

Above all, this situation reflects a lack of knowledge. Very few shipwrecks have been excavated in the Eastern Mediterranean, since they are much more numerous in the Western part of the sea.[19] The difference between the Eastern and Western waters was more pronounced in the Hellenistic period.[20] During the last few decades, when conferences about Hellenistic economies were held and when ship experts wrote scholarly studies about navigation in the Eastern Mediterranean, almost no underwater remains of this period had been discovered in that part of the sea. But during last decade, the situation has changed slightly, because more shipwrecks have been discovered in the Eastern Mediterranean. Publications about Hellenistic underwater discoveries off the Eastern Mediterranean coasts are therefore more numerous.[21]

These new discoveries could have been published in the Tropis conference proceedings, organized by Harry Tzalas of the Hellenic Institute for the Preservation of Nautical Tradition.[22] Nonetheless, even without them, these proceedings gather together many useful interdisciplinary studies about ships, relying on written and

---

17 Casson 1971 (2nd ed. 1995); Velissaropoulos 1980, who focuses mainly on maritime law; Basch 1987, who analyses the iconography to understand the architectural features of the ships; Murray 2012, who deals with military ships.
18 Pomey and Tchernia 1978; Gianfrotta and Pomey 1981; Pomey 1997.
19 For a discussion about the reasons, see Gianfrotta and Pomey 1981, 55–60; Parker 1992; Arnaud 2013, 199–200; Nantet 2016, 251–54. Also see chapter 4 in this volume.
20 See chapter 5.
21 In Israel, see the research of Jacob Sharbit and especially Ehud Galili: Galili et al. 2010; Syon et al. 2013; Galili et al. 2016a; Galili et al. 2016b. In Greece, see the survey led off Chios and Kythnos, Sakellariou et al. 2007. In Cyprus, see Demesticha 2011. In Egypt, see the study of the shipwreck of Heracleion (cf. chapter 5).
22 Tzalas 1989, 1990, 1995, 1996, 1999, 2001, 2002a, 2002b.

archaeological sources.[23] Unfortunately, the findings of the last two international meetings, held in 2005 in Ayia Napa (Tropis IX) and Hydra (Tropis X), have not been published so far and no other conference has been organized since then.[24] Since the end of the Tropis conferences, the publications have scattered in national periodicals. Thus, ship archaeologists have lost a place to meet and discuss the issue transnationally. But the revival in this field seems to come from Cyprus. The island is located in an appropriate place, central enough in the Eastern Mediterranean to facilitate these international meetings. It has peaceful relationships with its surrounding neighbours. In addition, the Archaeological Research Unit of the University of Cyprus played a major role in setting up maritime conferences,[25] some of them attaching much importance to ships. For instance, the workshop held in Cyprus, which resulted in this book, was intended to fulfil a need among the ship archaeologists interested in the Eastern Mediterranean to meet and discuss their research. The Honor Frost Foundation conference, 'Mediterranean Maritime Archaeology: Under the Mediterranean', held in Nicosia in October 2017, also offered a place of exchange for the scholars involved in this field of research.[26] Although, for the past decade, the foundation has been subsidising much work by scholars from all the countries of the Eastern Mediterranean, it has recently decided to restrict its funds to Cyprus, Egypt, Lebanon, and Syria. The strong recent revival of maritime archaeology in the Eastern Mediterranean is well evidenced by a recent book by Justin Leidwanger, which focuses on the Roman period.[27]

Unfortunately, the subject areas are overly compartmentalized. The ship archaeologists and the experts in ancient economics organise their own conferences. These barriers explain why ships have often been neglected in previous studies of Hellenistic economies.

---

23   Tzalas 2019.
24   Tropis IX. 9th International Symposium on Ship Construction in Antiquity, Ayia Napa 2005. Tropis X. 10th International Symposium on Ship Construction in Antiquity, Hydra 2008.
25   Among many meetings, one example: 'Per Terram, Per Mare. Production and Transport of Roman Amphorae in the Eastern Mediterranean', Nicosia, Cyprus, from 12 to 15 April 2013, organized by S. Demesticha, A. Kaldeli, D. Michaelides and V. Kassianidou.
26   Blue 2019.
27   Leidwanger 2020.

To meet this need, the studies gathered in this book aim to shed light on navigation in the Eastern Mediterranean during the Hellenistic period. They deal with all the parts of the Eastern Mediterranean: not only the Aegean Sea, but other seas between Asia Minor and Egypt. They use all kinds of data — literary, epigraphical, papyrological, iconographic and archaeological. The goal of the book is not to give an exhaustive analysis of maritime commerce; it is to set up the initial framework to help future scholars in their research, as more and more archaeological shipwrecks continue to be discovered and made public in the next decades.

This chapter is followed by a study, conducted by Jean-Marie Kowalski, about the role of Cyprus in the network of maritime routes (chapter 2). He demonstrates the differences between the distances given by the literary sources. The next chapter, written by Patrice Pomey, deals with the architectural type of the Hellenistic ships (chapter 3). He shows that the main change in ship evolution was the adoption of the tenon-and-mortise assemblage. The warships, once they were built, were decorated by ship painters (chapter 4). The Hellenistic period saw an increase in the tonnage (chapter 5). The comparisons between the epigraphical, papyrological and archaeological sources allow us to understand the chronological phases of this increase, and the factors that affected it. The last chapter is dedicated to Ship 17 from Thonis-Heracleion (chapter 6). The careful analysis of the architectural features by Alexander Belov shows that this ship was an Egyptian *baris* and that she may have sailed in the estuary of the Nile.

Bibliography

Archibald, Z. H., J. K. Davies, V. Gabrielsen, and G. J. Oliver, eds. 2001. *Hellenistic Economies*. London: Routledge. https://doi.org/10.4324/9780203995921

Archibald, Z. H., J. K. Davies, and V. Gabrielsen, eds. 2005. *Making, Moving and Managing: The New World of Ancient Economies, 323–31 B.C.* Oxford: Oxbow Books. https://doi.org/10.1111/j.1468-0289.2006.00361_13.x

Archibald, Z. H., J. K. Davies, and V. Gabrielsen, eds. 2011. *The Economies of Hellenistic Societies, Third to First Centuries BC*. Oxford: Oxford University Press. https://doi.org/10.1093/acprof:osobl/9780199587926.001.0001

Arnaud, P. 2013. 'L'apport de l'archéologie sous-marine.' In *Archéologie sous-marine: pratiques, patrimoine, médiation*, edited by C. Cérino, M. L'Hour, and É. Rieth, 193–203. Rennes: Presses Universitaires de Rennes.

Arnaud, P. 2015. 'La batellerie de fret nilotique d'après la documentation papyrologique (300 avant J.-C.–400 après J.-C.).' In *La Batellerie égyptienne: archéologie, histoire, ethnographie*, edited by P. Pomey, 99–150. Alexandria: Centre d'Études Alexandrines.

Basch, L. 1987. *Le Musée imaginaire de la marine antique*. Athens: Institut hellénique pour la préservation de la tradition nautique.

Blue, L., ed. 2019. *In the Footsteps of Honor Frost. The Life and Legacy of a Pioneer in Maritime Archaeology*. Leiden: Sidestone Press.

Bresson, A. 2011. 'Grain from Cyrene.' In *The Economies of Hellenistic Societies, Third to First Centuries BC*, edited by Z. H. Archibald, J. K. Davies, and V. Gabrielsen, 66–95. Oxford: Oxford University Press. https://doi.org/10.1093/acprof:osobl/9780199587926.003.0004

Bresson, A. 2009. *L'Économie de la Grèce des cités (fin VIe–Ier siècle a.C.). I, Les structures et la production*. Paris: Armand Colin. https://doi.org/10.14375/np.9782200265045

Bresson, A. 2018a. 'Flexible Interfaces of the Ancient Mediterranean World.' In *Trade and Colonization in the Ancient Western Mediterranean: The Emporion, From the Archaic to the Hellenistic Period*, edited by E. Gailledrat, M. Dietler, and R. Plana-Mallart, 35–46. Montpellier: Presses Universitaires de la Méditerranée (Collection *Monde Ancien*).

Bresson, A. 2018b. 'Coins and Trade in Hellenistic Asia Minor: The Pamphylian Hub.' In *Infrastructure and Distribution in Ancient Economies Proceedings of a Conference Held at the Austrian Academy of Sciences, 28–31 October 2014*, edited by B. Woytek, 35–46. Vienna: Austrian Academy of Sciences. https://doi.org/10.2307/j.ctvddzgz9.9

Casson, L. 1971. *Ships and Seamanship in the Ancient World*. 2nd ed. Princeton: Princeton University Press.

Chankowski, V. and F. Duyrat, eds. 2004. *Le Roi et l'économie: autonomies locales et structures royales dans l'économie de l'empire séleucide: actes des rencontres de Lille (23 juin 2003) et d'Orléans (29–30 janvier 2004)*. Lyon: Maison de l'Orient méditerranéen.

Demesticha, S. 2011. 'The 4th-Century-BC Mazotos Shipwreck, Cyprus: A Preliminary Report.' *International Journal of Nautical Archaeology* 40: 39–59. https://doi.org/10.1111/j.1095-9270.2010.00269.x

Finley, M. I. 1985. *The Ancient Economy*. London: Hogarth.

Galili, E., B. Rosen, and A. Zemer. 2016a. 'A Marble Disc (Ship Eye) from a Hellenistic Shipwreck South of Haifa.' In *Seafarers' Rituals in Ancient Times*, edited by A. Zemer, 130–31. Haifa: The National Maritime Museum.

Galili, E., D. Syon, G. Finkielsztejn, V. Sussman, and G. D. Stiebel. 2016b. 'Late Ptolemaic Assemblages of Metal Artifacts and Bronze Coins Recovered off the Coast of Atlit.' *Atiqot* 87: 135.

Galili, E., V. Sussman, G. D. Stiebel, and B. Rosen. 2010. 'A Hellenistic/Early Roman Shipwreck Assemblage off Ashkelon, Israel.' *International Journal of Nautical Archaeology* 39: 125–45. https://doi.org/10.1111/j.1095-9270.2009.00249.x

Gianfrotta, P. A. and P. Pomey. 1981. *Archeologia subacquea: storia, tecniche, scoperte e relitti*. Milan: A. Mondadori.

Gibbins, D. 2001. 'Shipwrecks and Hellenistic Trade.' In *Hellenistic Economies*, edited by Z. H. Archibald, J. K. Davies, V. Gabrielsen, and G. J. Oliver, 273–312. London: Routledge. https://doi.org/10.4324/9780203995921

Harris, E. M., D. M. Lewis, and M. Woolmer, eds. 2016. *The Ancient Greek Economy: Markets, Households and City-States*. Cambridge: Cambridge University Press. https://doi.org/10.1017/CBO9781139565530

Lawall, M. 2005. 'Amphoras and Hellenistic Economies: Addressing the (Over-)Emphasis on Stamped Amphora Handles.' In *Making, Moving and Managing: The New World of Ancient Economies, 323–31 B.C.*, edited by Z. H. Archibald, J. K. Davies and V. Gabrielsen, 188–232. Oxford: Oxbow Books.

Leidwanger, J. 2020. *Romans Seas. A Maritime Archaeology of Eastern Mediterranean Economies*. Oxford: Oxford University Press.

Murray, W. M. 2012. *The Age of Titans: The Rise and Fall of the Great Hellenistic Navies*. Oxford: Oxford University Press. https://doi.org/10.1093/acprof:oso/9780195388640.001.0001

Nantet, E. 2016. *Phortia: le tonnage des navires de commerce en Méditerranée: du VIIIe siècle av. l'ère chrétienne au VIIe siècle de l'ère chrétienne*. Rennes: Presses Universitaires de Rennes.

Pomey, P., ed. 1997. *La Navigation dans l'Antiquité*. Aix-en-Provence: Édisud.

Pomey, P. and A. Tchernia. 1978. 'Le tonnage maximum des navires de commerce romains.' *Archaeonautica* 2: 233–51.

Préaux, C. 1939. *L'Économie royale des Lagides*. Bruxelles: Éd. de la Fondation égyptologique reine Élisabeth.

Rostovtzeff, M. I. 1941. *The Social & Economic History of the Hellenistic World*. Oxford: The Clarendon Press.

Sakellariou, D., P. Georgiou, A. Mallios, V. Kapsimalis, D. Kourkoumelis, P. Micha, T. Theodoulou, and K. Dellaporta. 2007. 'Searching for Ancient Shipwrecks in the Aegean Sea: The Discovery of Chios and Kythnos Hellenistic Wrecks with the Use of Marine Geological-Geophysical Methods.' *International Journal of Nautical Archaeology* 36: 365–81. https://doi.org/10.1111/j.1095-9270.2006.00133.x

Scheidel, W., I. Morris, and R. P. Saller, eds. 2008. *The Cambridge Economic History of the Greco-Roman World*. Cambridge: Cambridge University Press. https://doi.org/10.1017/chol9780521780537

Syon, D., C. Lorber, and E. Galili. 2013. 'Underwater Ptolemaic Coin Hoards from Megadim.' *Atiqot* 74: 1–8.

Tzalas, H., ed. 1989. Tropis I: 1st International Symposium on Ship Construction in Antiquity: proceedings, Piraeus, 1985. Athens: Hellenic Institute for the Preservation of Nautical Tradition.

Tzalas, H., ed. 1990. Tropis II: 2nd International Symposium on Ship Construction in Antiquity: proceedings, Delphi, 1987. Athens: Hellenic Institute for the Preservation of Nautical Tradition.

Tzalas, H., ed. 1995. Tropis III: 3rd International Symposium on Ship Construction in Antiquity: proceedings, Athens, 1989. Athens: Hellenic Institute for the Preservation of Nautical Tradition.

Tzalas, H., ed. 1996. Tropis IV: 4th International Symposium on Ship Construction in Antiquity, Center for the Acropolis Studies, Athens, 28, 29, 30, 31 August 1991: proceedings. Athens: Hellenic Institute for the Preservation of Nautical Tradition.

Tzalas, H., ed. 1999. Tropis V: 5th International Symposium on Ship Construction in Antiquity: proceedings, Nauplia, 26, 27, 28 August 1993. Athens: Hellenic Institute for the Preservation of Nautical Tradition.

Tzalas, H., ed. 2001. Tropis VI: 6th international symposium on ship construction in antiquity Lamia, 28,29,30 August 1996: proceedings. Athens: Hellenic Institute for the Preservation of Nautical Tradition.

Tzalas, H., ed. 2002a. Tropis VII: 7th International Symposium on Ship Construction in Antiquity, Pylos, 26, 27, 28, 29 August 1999: proceedings. Athens: Hellenic Institute for the Preservation of Nautical Tradition.

Tzalas, H., ed. 2002b. Tropis VIII: 8th International Symposium on Ship Construction in Antiquity: Hydra, 27, 28, 29, 30 august 2002: proceedings. Athens: Hellenic Institute for the Preservation of Nautical Tradition.

Tzalas, H. 2019. 1985-2008: TROPIS International Symposia on Ship Construction in Antiquity. In *In the Footsteps of Honor Frost. The Life and Legacy of a Pioneer in Maritime Archaeology*, edited by L. Blue, 91–94. Leiden: Sidestone Press.

Velissaropoulos, J. 1980. *Les Nauclères grecs: recherches sur les institutions maritimes en Grèce et dans l'Orient hellénisé*. Geneva: Droz.

2. Evolutions of the Representation of the Eastern Mediterranean in the Hellenistic Period

Jean-Marie Kowalski

Studying the evolutions of the representation of the Eastern Mediterranean in the Hellenistic period is a quite challenging task. There is a substantial lack of evidence and in some cases, the granularity of the manuscript sources is quite poor. Moreover, the quality of information varies from one source to another. Nevertheless, by comparing these sources we create a context within which to examine the different ways Cyprus was integrated within the network of maritime routes. The calculation of the length of these routes is unsurprisingly based on an asymmetrical representation of space, but a closer look reveals the importance of seasonality to navigation. The winds blow from sharply different directions in summer and in winter, so it was sometimes impossible to sail certain routes, and a statistical assessment demonstrates that some were much more frequently sailed in summertime than in winter. Lastly, the variety of landscapes around Cyprus makes it necessary to focus on the different kinds of landmarks, and what makes them products of their environment.

This chapter will address the representation of the Eastern Mediterranean during the Hellenistic period, focusing in particular on the case of Cyprus and its place within the maritime routes of the Eastern Mediterranean. I shall compare Classical and Ancient representations with late Hellenistic examples in order to highlight some of the differences and continuities of the representation of this particular region.

The first difficulty that arises is the lack of reliable, datable evidence from this period. Cyprus is very often mentioned in classical literature: for example in the writings of Herodotus, Xenophon, Plato, Thucydides, Isocrates, Demosthenes, and Ephorus.[1] Until the late first century BC (Diodorus Siculus, Strabo) and the first century AD (Flavius Josephus,[2] Dioscorides[3]) almost no information is given about the island and its geography apart from brief references in Menander's comedies, in Polybius' histories, in Theopompus' fragments or in those by Clearchus the philosopher (fourth century BC).[4] Most of the references deal with political or military issues, but very inadequate information is given about navigation and maritime routes.[5]

We will pay a particular attention to the *Stadiasmus* (or *Periplus Maris Magni*) and to Strabo's *Geography*, even if the *Stadiasmus* raises several challenging questions as it is rather difficult to say precisely when it was written and the author's sources.[6] Timosthenes of Rhodes (c. 270 BC) is one of the most important sources, but some late information dating from 10 BC can also be identified. It is even more difficult to say when this book was written as the different assumptions range from 50 CE

---

1 Hdt., 1.72.10; 1.105.10; 1.199.26; 2.79.4; 3.91.7; 4.162.7; 4.164.6; 5.31.14; 5.49.30; 5.108.7; 5.109.2; 5.109.10: 5.113.12; 5.115.3. Xenophon, *Hellenica* 2.1.29; 4.8.24; 5.1.10; 5.1.31; *Cyropaedia* 8.6.8; 8.8.1; Athenaeus, *Resp.* 2.7.4. Plato, *Menexenus* 241e. Thucydides, 1.94.2; 1.104.2;1.112.2; 1.112.4; 1.128.5. Isocrates, *Panegyricus* 134.7; 141.2; 153.6; 161.2; *Evagoras* 18.5; 51.4; 53.1; 58.2; 60.2; 62.3; 67.5; *de Pace* 86.5; *Philippus* 62.3; 102.1; Demosthenes, *adversus Leptinem* 76.4. Ephorus, Jacoby *FGrH*, 70F119; 70F191; 70F192; 70T20. All the sources used here are issued from the Loeb edition, unless mentioned otherwise.

2 *Antiquitates Judicae* 1.128; 13.132; 13.285; 13.287; 13.328; 13.331; 13.358; 14.121; 15.184; 16.144; 16.197; 18.138; 18.130; 18.131; 18.138; 18.148; 18.159; 18.160; 20.52; *Contra Apionem* 1.99; *de Bello Judaico* 1.181; 1.407; 1.418; 2.108; 2.220; 4.469.

3 *De materia medica* 1.71.1; 1.97.4; 1.127.3; 5.32.1; 5.76.2; 5.91.1; 5.102.1; 5.103.1; 5.109.1; 5.138.1.

4 Menander, *Misumenus* 32; 231, Fr. 5 l.1; Fr. 151 l. 231; Polybius, *Historiae* 5.34.7; 5.59.5; 12.25f.2; 18.54.1; 18.55.6; 27.13.1; 29.27.9; 33.5.1; 31.10.3–10; 31.17.4–8; 31.18.2–8; 31.20.6; 33.5.2; 33.11.7; 39.7.6; Theopompus, Jacoby, *FGrH* 115F19, 103, 105, 116, 222; Clearchus Phil., *Fragmenta*, Fr. 19; 43a.

5 Other sources are only fragmentary or completely lost, such as Artemidorus' description of earth (first century BCE); Posidonius (from second to the first century BCE); and Timosthenes' *About Harbours* (from the second half of the third century BCE), who influenced Strabo, Eratosthenes and Dicearchus.

6 See Arnaud, P. 2009. 'Notes sur le *Stadiasme de la Grande Mer*: la Lycie et la Carie.' *Geographia Antiqua* 18.165–193.

to the fifth century. There are several layers of information that mainly belong to the Hellenistic period and the beginning of the Roman era, but they are probably scattered on a very large span of time. It is also important to note that this text was written long after the Hellenistic period and Strabo's *Geography*. We must therefore recognise that this document cannot be considered a fully reliable piece of evidence that reflects the way maritime spaces were represented during a specific period. What is more, there is no real consistency in the descriptions of the different geographical areas within the *Stadiasmus*. The author gives a very accurate depiction of the coast on the west of Alexandria — this is the only part of the text that could be compared to modern Nautical Instructions — but there is a real lack of detail in the description of the coast of Asia Minor.

As far as the Archaic period is concerned, the island of Cyprus is mentioned only once in Homer's *Iliad*,[7] and five times in the *Odyssey*.[8] In the *Iliad*, Cinyras learns that the Achaeans are about to sail to Cyprus. This rumour appears to spread beyond the Aegean, but absolutely no indication is given about the island itself. We therefore cannot really say that Cyprus was integrated in the maritime communications network during the poet's time. In the *Odyssey*, Cyprus is first mentioned as the place visited by Menelaus when he is wandering on his way back to his country. He then calls in at Phoenicia, in Egypt, before meeting the Ethiopians, the Sidonians, the Eremboï and the Libyans in a clockwise trip around the Eastern Mediterranean. In Cyprus, one can also see Aphrodite's forest and altars in Paphos.[9] Ulysses also arrives in Cyprus after he has been captured as a pirate in Egypt.[10]

We cannot draw any significant conclusions from Homer's references to the island of Cyprus, apart from the fact that the island was clearly well integrated into the network of maritime routes that criss-crossed the Eastern Mediterranean and the Aegean Sea. No information is given about its harbours and ports, although the sanctuary of Aphrodite in Paphos appears to have been considered a useful landmark for sailors. The existence of this sanctuary is the only accurate information about Cyprus given in Homer's *Odyssey*, and it is also mentioned in the

---

7   *Il.* 11.21.
8   *Od.* 4.83; 8.362; 17.443–444; 17.448.
9   *Od.* 8.362.
10 *Od.* 17.443; 448.

Stadiasmus,[11] although it is said to be situated near a city that faces the south, with a triple harbour. Homer's geographical knowledge about Cyprus was evidently quite poor.

While the poet describes Paphos, Aphrodite's birthplace,[12] possessing an altar dedicated to the goddess as well as a forest, the *Stadiasmus*[13] does not include the forest — although the sanctuary is still there, as is a south-facing city with a triple harbour, which is said to be accessible in all wind conditions since its entrance faces the south. This is the only 'triple harbour' mentioned in the *Stadiasmus* with breakwaters that were built during the Hellenistic period.[14] It is therefore a particularly interesting harbour, because it is highly representative of the Hellenistic world before the damage wreaked by earthquakes and by the constant silting of the basins.[15] The exact meaning of 'triple harbour' is not completely clear, but it is highly probable that this refers to the division of the main basin into several parts, inside a *limen kleistos*, a closed harbour. This naming is specifically Hellenistic, as 'closed harbours' are not mentioned in classical literature, but they are present in Hellenistic writings.

2.1. Granularity of Information

The granularity of the information given is rather different in the *Stadiasmus* and in Strabo's *Geography* (Figs. 2.1 and 2.2).

---

11 297.1.
12 *Od.* 8.364.
13 297
14 Raban 1995, 168, Fig. 42. Paphos 2: See Leonard, Dunn and Hohlfelder 1998, 151, Fig. 4.
15 The shape of the triple harbour described in the *Stadiasmos* has been imagined in different ways. For instance, it has been suggested that there was a triple internal division with the main basin contained inside the breakwaters, and at the same time the use of the bays to the north and south. Geophysical surveys have revealed that the bedrock of the basin is divided into two uneven parts, upon which can be identified the remains of constructions that in effect would have created two basins. The placement of a wharf in the Western part of the basin could in theory have created a harbour with three sections. Surveyors' plans reveal remnants of building material at two points at right angles to the beach in the west harbour. A triple scheme could also be envisioned using the natural separation of the Eastern harbour from the stream that flowed into it. Similarly, the triple harbour may have consisted of the division of the Eastern and Western sections of the port, and also utilised the natural bay that exists to the south, which was used in medieval times when the main harbour became too silted.

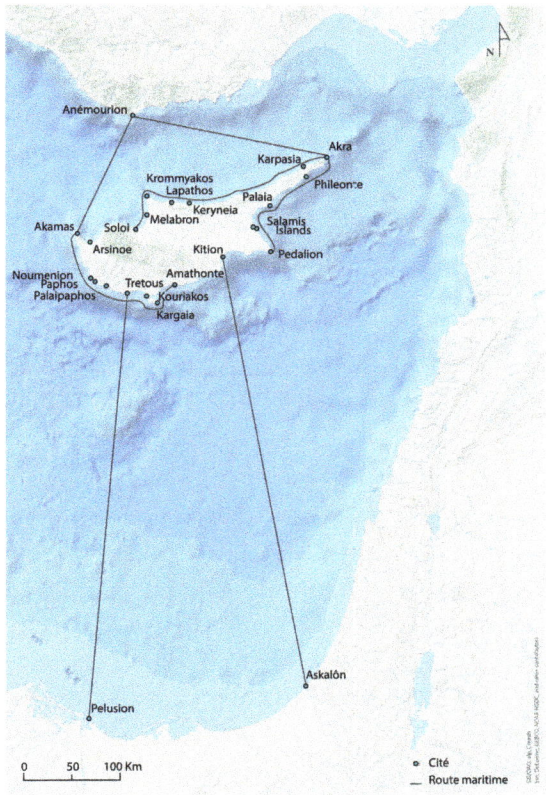

Fig. 2.1 Itineraries mentioned in the *Stadiasmus*
(CAD Anne-Laure Pharisien/CReAAH).

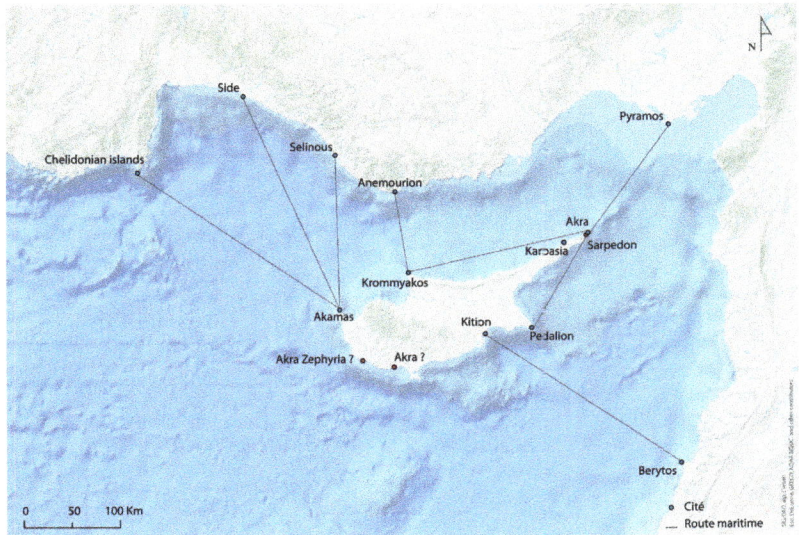

Fig. 2.2 Itineraries mentioned in Strabo's *Geography*
(CAD Anne-Laure Pharisien/CReAAH).

In the *Stadiasmus*, no fewer than twenty-two different itineraries from Cyprus or around the island are mentioned, while in the *Geography*, some thirteen routes around Cyprus, and between Cyprus and other places, are mentioned.[16] At a first glance, it looks as if Strabo's *Geography* and the *Stadiasmus* share many commonalities, but a further examination of these texts reveals significant discrepancies between them. Indeed, among the twenty-two different itineraries around and from Cyprus mentioned in the *Stadiasmus*, eighteen are parts of the periplus around the island, while Strabo does not give very accurate information about the distances around Cyprus, but he replaces it in the global network of maritime routes. That is to say that Strabo's homage to Homer in his introduction to the *Geography* is not some kind of compulsory tribute,[17] but a true allegiance to Homer's vision of the world. Strabo's representation places Cyprus inside a network of maritime routes[18] while the *Stadiasmus* focuses the reader's attention on a large number of sometimes very short itineraries around the island. It looks as if the discrepancy between testimonies was not a matter of their age, but rather a matter of sources and purposes. The *Stadiasmus* also reveals the very dense network of harbours and port facilities that seafarers could find around the island.

In spite of these discrepancies, some common features can be identified between these representations.

2.2. Distances and Maritime Experience

At first it might appear both difficult and almost meaningless to make comparisons between the distances mentioned by the *Stadiasmus* and those mentioned by Strabo, as the former mainly deals with short-range itineraries while the latter deals with long-range journeys. However, if we consider the quantitative information given by the authors, the *Stadiasmus* and the *Geography* give similar information about the distances involved. Nonetheless, no firm conclusions can be drawn from the distances mentioned, as many different factors have led to some irrelevant indications. The main factor is the granularity of manuscripts.

---

16 See Figs. 2.1 and 2.2.
17 *Geography* 1.1.2.
18 He also mentions distances within the island.

The two main editions in use nowadays derive from one tenth-century manuscript (*Matritensis* 121), in which the text of the *Stadiasmus* comes immediately after the *Chronicle* of Hippolytus.[19] This unique manuscript is badly damaged and can hardly be deciphered. That is why Müller's edition of the *Stadiasmus* in the *Geographi Graeci Minores* contains many corrections and much additional information. Helm's edition is more recent (1929) but it does not contain as many corrections of Müller's edition. The manuscript history of the *Stadiasmus* means that the reader must be very careful when applying the quantitative data given by the text about the different itineraries.

These factors will not be thoroughly discussed, but one can see, for example, that according to the *Stadiasmus*, Kargaia is supposedly only 40 stadia[20] away from Kouriakos, while the true distance is approximately 13 nautical miles. This would suggest that there are only 3.08 stadia per mile, but the distance between Keryneia and Lapathos is said to be 450 stadia: it is in fact no more than 6 miles.

Whatever the causes of these discrepancies, they should make one very cautious when assessing the reliability of the distances given. Nevertheless, the average number of stadia per mile is very similar between Strabo and the *Stadiasmus*: 12.7 for Strabo and 14.3 for the *Stadiasmus*.

Even if each individual indication cannot be considered fully reliable, these are quite close as a group, in spite of differences between the nature of sailing as outlined in these texts. While Strabo mainly refers to long-distance routes on the high seas, the *Stadiasmus* merges different types of journey, from very short coastal navigation to oceangoing maritime routes. Therefore, the apparent resemblance of their representations is a kind of *trompe l'œil* similarity insofar as it is not based on the same items. Additionally, nothing is said about the size and type of ships that are supposed to sail these routes.

Some indications about the weather conditions given by the *Stadiasmus* shed new light on these distances, since they introduce qualitative features to the long lists of distances. Paphos[21] is said to be a triple harbour whatever the wind conditions are, just like the city of

---

19   See Arnaud 2009.
20   There are almost 79 stadia per nautical mile.
21   297.1.

Ammochostos.[22] On the contrary, Amathus[23] is said to be deprived of any kind of harbour (*alimenos*) which makes it an unsafe destination. But this does not mean that the city does not have any mooring place. Strabo says that Amathus is a city but does not say anything about its facilities. The indication of strong gales that blow from the north (*boreas* wind)[24] in Arsinoe and in Karpaseia[25] raises the question of the precise meaning of the verb used by Strabo. The author writes '*kheimazei*', which can be understood in two different ways, as '*cheimazein*' refers to winter conditions rather than to generic storms. On the same coast, Melabron[26] is said to have good summer mooring.

2.3. Seasonality of Weather Indications

At first glance, the *Stadiasmus* does not indicate directions in the same way in all descriptions. As far as Cyprus is concerned, directions are described using cardinal points to indicate wind directions. One specific type of wind is mentioned, the *zephyros*, which blows from the west, while the south is said to be '*mesembria*', that is to say, from the sun's side at noon.

The contemporary weather statistics of this area provided by the National Oceanic and Atmospheric Administration (NOAA) provide useful indications. Indeed, during the summer, winds usually blow from the west, while in winter, they predominantly blow from the north east with an average speed of 10 knots.[27] When they blow from the north, they are usually stronger and reach an average speed of 15 knots. In both cases, they make harbours and moorings quite difficult and unsafe for ships, as even moderate winds from the north east usually become stronger along the northern coast.

This seems to be a significant indication of the seasonality of navigation or, at least, the seasonality of distance indications, as the *Stadiasmus* explicitly mentions unfavourable winter conditions on the

---

22 304.1.
23 302.1.
24 309.1.
25 314.1.
26 310.1.
27 See https://opencpn.org/OpenCPN/info/downloadplugins.html . These indications do not take into account local winds such as sea breezes, land breezes, venturi effects in narrow places or modifications of wind direction around capes.

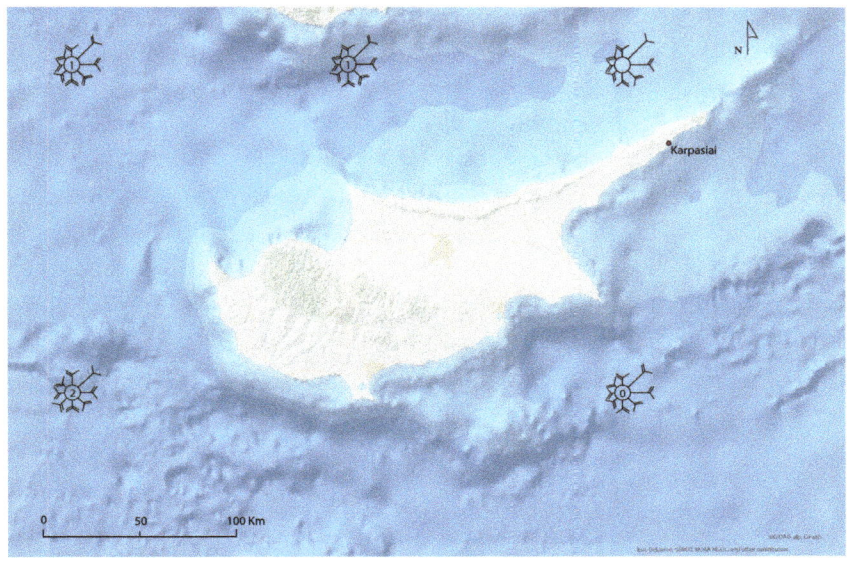

Fig. 2.3 Weather conditions around Cyprus in December
(CAD Anne-Laure Pharisien/CReAAH).

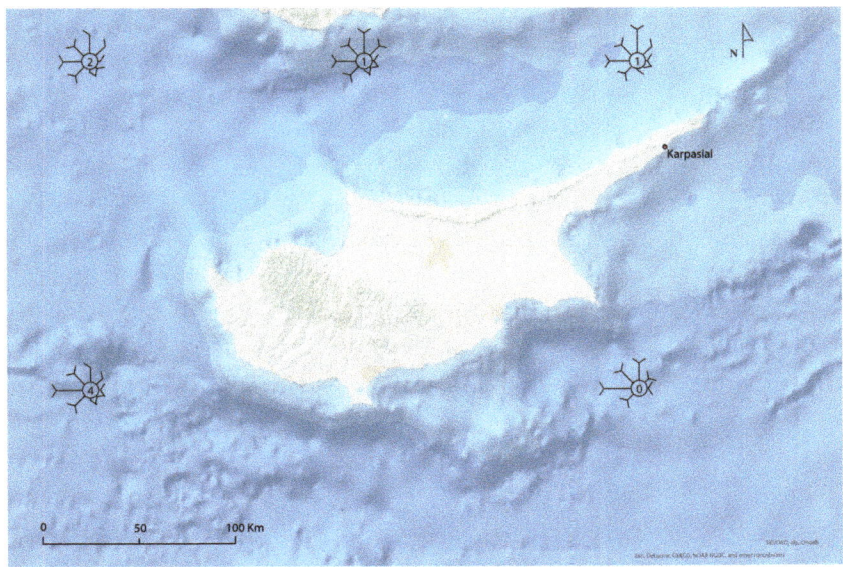

Fig. 2.4 Weather conditions around Cyprus in June
(CAD Anne-Laure Pharisien/CReAAH).

north coast of Cyprus and does not deal exclusively with summer navigations (Figs. 2.3 and 2.4).

In winter, the average north wind is stronger than the average northeasterly wind. Lastly, these indications do not take into account the local weather phenomena that were well known in antiquity.

This qualitative information suggests a new approach to the indications of distances from and around the island. Instead of paying attention solely to these distances, the information given about the quality of harbours and moorings suggest that we should make connections between weather conditions and distances.

2.4. Influence of Weather Conditions Over Navigation

Even if climatology has changed within a timespan of two millennia, the lack of statistics before the second half of the twentieth century made it acceptable to use NOAA's data. We have decided here (table 2.1) to rate this data according to the average winter (end of December) and summer conditions (end of June) according to the angle between the wind and the route supposedly followed by the ships along the itineraries mentioned (from 180 to 135 degrees: 3 — fair conditions –, from 135 to 90 degrees: 4 — highly favourable conditions –, around 90 degrees: 2 — average conditions –, from 90 to 45 degrees: 1 — poor conditions –, and from 45 to 0 degrees: 0 — unfavourable conditions –).

If this assessment can be applied to the capacities of the ancient ships, two further conclusions can be drawn. First, Strabo and the author of the *Stadiasmus* both mention maritime itineraries that are more favourable for ships during summer. This does not mean that sailing in winter was impossible, but the winds were much less favourable, and some harbours and moorings were made unsafe, especially on the north coast of Cyprus. Secondly, even if the *Stadiasmus* and Strabo both rely on information derived from accounts of summer navigation, the *Stadiasmus* seems to depend more explicitly on the maritime experience of sailors, because the itineraries mentioned are more favourable during summer.

2. Representation of the Eastern Mediterranean

Table 2.1 Comparison of the impact of weather conditions on navigation.

Source	Departure	Arrival	Favorable wind in summer 0=poor 4=high	Wind	Favorable wind in winter 0=poor 4=high	Wind
Stadiasmus	Kouriakos	Kargaia	4,00	W 10	0	variable 15
Stadiasmus	Noumenion	Palaipaphos	4,00	W 10	0	variable 15
Stadiasmus	Palaipaphos	Tretous	4,00	W 10	0	variable 15
Stadiasmus	Pedalion	Islands	0,00	W-N 10	2	variable NE 15
Stadiasmus	Lapathos	Karpaseia	4,00	W 10	0	NE 15
Stadiasmus	Kouriakos	Amathus	3,00	W 10	0	variable NE 15
Stadiasmus	Krommyakos	Melabron	3,00	W 10	3	variable NE 15
Stadiasmus	Paphos	Noumenion	4,00	W 10	2	variable NE 15
Strabo	Krommyon	Kleides Islands	3,00	W-N 10	0	NE 15
Stadiasmus	Karpaseia	Akra	3,00	W-N 10	0	NE 15
Stadiasmus	Soloi	Keryneia	3,00	W-N 10	0	NE 15
Stadiasmus	Salamis	Palaia	3,00	W-NW 10	0	NE 15
Strabo	Anemourion	Krommyon	4,00	W-N 10	2	E-NE 15
Stadiasmus	Palaia	Phileonte	3,00	W-N 10	0	NE 15
Stadiasmus	Islands	Salamis	1,00	W-N 10	3	NE 15
Stadiasmus	Akamas	Paphos	4,00	W-N 10	3	NE 10-15
Strabo	Kleides Islands	Pedalion	3,00	W-N 10	3	NE 15
Strabo	Kleides Islands	Pyramos	2,00	W-N 10	1	NE 15
Stadiasmus	Anemourion	Akamas	2,00	W-N 10	3	NE 15
Stadiasmus	Phileonte	Akra	2,00	W-N 10	0	NE 15
Strabo	Kition	Berytos	4,00	W-NW 10	0	NE-E 15
Strabo	Akamas	Side	1,00	W-N 10	2	NE 10
Strabo	Akamas	Selinus	1,00	W-N 10	1	NE 10
Strabo	Akamas	Chelidonian Islands	0,00	W-N 10	3	NE 10
Stadiasmus	Melabron	Soloi	2,00	W-N 10	3	NE 15
Stadiasmus	Akamas	Arsinoe	4,00	W-N 10	3	NE 10
Stadiasmus	Keryneia	Lapathos	0,00	W-N 10	4	NE 15
Stadiasmus	Tretous	Kouriakos	4,00		0	
Stadiasmus	Akra	Anemourion	0,00		4	
		Strabo	**2,25**		**1,50**	
		Stadiasmus	**2,71**		**1,43**	

2.5. Granularity and Quality of Information: The Problem of Salience

Distance is a core datum in geographic literature, but both authors give a series of additional information that is rather qualitative than quantitative, and therefore shed a new light on the question of the evolution of the representation of maritime spaces. At a first glance, the granularity seems to be finer in the *Stadiasmus* than in Strabo's *Geography*.

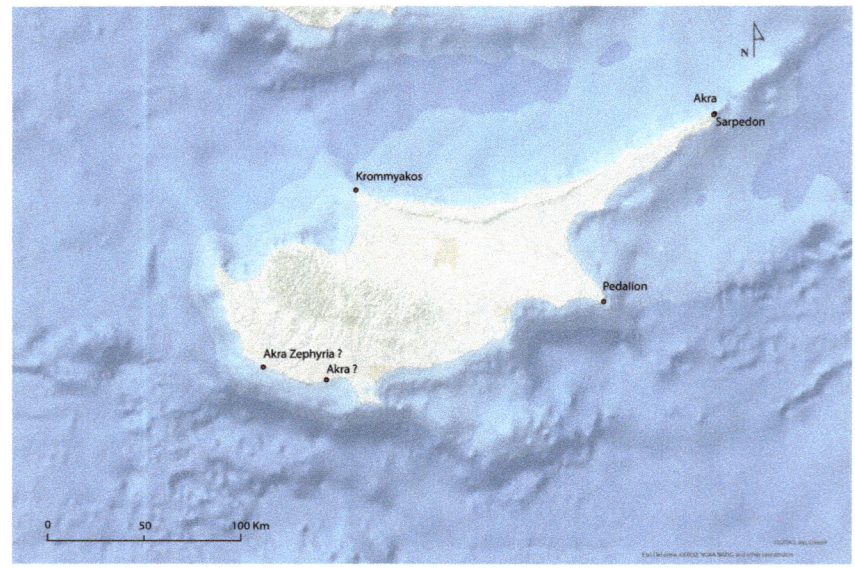

Fig. 2.5 *Akrai* and *akroteria* in Strabo's *Geography*
(CAD Anne-Laure Pharisien/CReAAH).

Fig. 2.6 *Akrai* and *akroteria* in the *Stadiasmus*
(CAD Anne-Laure Pharisien/CReAAH).

Some geographic entities, usually described as capes or promontories, should be considered as structural elements of space, especially the *akroteria* and *akrai* around Cyprus (Figs. 2.5, 2.6 and Table 2.2). *Akroteria* and *akrai* cannot be considered only as capes or promontories, that is to say horizontal or vertical salient geographical features. These are elements that organize space because they are useful landmarks which form nodes on the network of maritime routes.

Some of these *akrai* or *akroteria* can hardly be considered as visually salient landmarks, but they are undoubtedly cognitive landmarks. This is particularly the case for Akroterion Tretous in the *Stadiasmus*, which could be the *akra* Strabo places after Kourion. This place was known as the anchorage of al-Itritus during the Ottoman period.[28] This anchorage offered good protection against the winds blowing from the north (*boreas*) and the east/south east (*euros*).

This example highlights that geographical entities do have formal features, but they are also characterized by their multiple capabilities: to protect ships, to offer safe mooring places, to be good landmarks, to create a landing place on shore, to provide ships with fresh water, or to make seafaring possible. This could be the case with Tretous, as the adjective '*tretos*' generally describes rocks with holes through which mooring lines are to be pushed.[29]

Table 2.2 Main *akroteria* and *akrai* mentioned in the *Stadiasmus* and in Strabo's *Geography*.

Stadiasmus	Strabo
Akroterion Tretous (hardly located)[1]	Krommyon (cape Kormakitis) akroterion defined as an akra[2]
Akroterion Kargaia[3] (cape Gata): provides a harbour, a mooring place and water.	Anemourion (akra of Cilicia)[4]
Akra (cape Apostolos Andreas)[5]	Sarpedon akra (cape Aspostolos Andreas)[6]
	Akra (*kai oros*): cape of Aphrodite's temple[7].
	Akra Pedalion (cape Greco)[8]
	Akra after Kourion (no name = Tretous?)[9]
	Akra Zephyria (north of Palaipaphos)[10]

[1] 301.1.   [3] 303.1.   [5] 315.1.   [7] *Geogr.*, 14.6.3.   [9] *Geogr.*, 14.6.3.
[2] *Geogr.*, 14.5.3.   [4] *Geogr.*, 14.5.3.   [6] *Geogr.*, 14.5.4.   [8] *Geogr.*, 14.6.3.   [10] *Geogr.*, 14.6.3.

---

28   Rapoport and Savage-Smith 2014, 476.
29   See Homer, *Od.* 13.77; Dionysius, *Geogr., Per Bosporum navigatio* 47.4.

2.6. Salience and Visually Distinctive Features: The Case of Cape Pedalion

Salience derives from affordances, but salience also derives from visually distinctive features. The *Stadiasmus* description of Cyprus gives quite scanty details about these elements. Absolutely no information is given about Cape Pedalion, which is nowadays known as Cape Greco in the southern part of the island. On the other hand, Strabo's description provides the reader with accurate and granular details.[30] According to the geographer, Pedalion is a cape (*akra*), with a rough hill (*trakhus lophos*) on the top, which is high, and is both table-shaped and dedicated to Aphrodite. All these details give visual indications that characterize this cape and help the reader recognize it.

The concept of salience is also a relative one. Cyprus is an island with very sharp geographic contrasts, the Eastern part being much lower than the Western. In spite of this, Strabo's description of the Eastern cape of Cyprus mentions an *akra* with an *oros*, on the top of which (*akrôreia*) is built a temple to Aphrodite Akraia, which cannot be entered by women. In front of this cape lie several islands. The *Stadiasmus* does not mention any of these details, but what is striking here is that Strabo's description seems to describe a mountain in a location that is actually one of the lowest parts of Cyprus.

The only thing we can say is the little elevation at the very end of the cape is the only noticeable distinctive feature of this place. Therefore, the *oros* cannot be considered literally as a mountain, but simply as a prominent element in the landscape that characterizes an important landmark around Cyprus. What is more, this landmark is only noticeable when ships are navigating close to the coast, in an area made dangerous by the different islands around it.

2.7. Conclusion

Therefore, what makes the difference are actually the man-made buildings and facilities on shore. However, at the same time, the texts reveal that the basic features of the human representation of spaces

---

30 *Geographia* 14.6.3.

remain the same. While quantitative information cannot be considered as truly reliable for many reasons, such as the asymmetry of the distances estimated or some defects belonging to the manuscript, qualitative elements should be considered closely. Firstly, the texts we have studied put into sharp relief the fact that, if distances are usually based on rough estimates and closely linked to the length of time it takes to travel by sea, they are also somehow connected with seasonality and the estimation of good travel conditions. What is more, the definition of geographical elements at the end of the Hellenistic period is still based on affordances rather than on formal features. These geographers could say that a place was an *oros* even if it was not a mountain, just because it was somehow higher than its environment, and this specific feature made it highly salient in its environment. *Akraï* are neither particularly large capes nor high promontories, but they are salient in their environment and can provide seafarers with safe moorings, or offer good protection against the winds in specific conditions.

Bibliography

Arnaud, P. 2009. 'Notes sur le *Stadiasme de la Grande Mer*: la Lycie et la Carie.' *Geographia Antiqua* 18: 165–93.

Leonard, R. J., R. K. Dunn, and R. L. Hohlfelder. 1998. 'Geoarchaeological Investigations in Paphos Harbour, 1996.' *Report of the Department of Antiquities of Cyprus*, 141–57.

Raban, A. 1995. 'The Heritage of Ancient Harbour Engineering in Cyprus and the Levant.' In *Proceedings of the International Symposium Cyprus and the Sea*, edited by V. Karageorgis and D. Michaelides, 139–90. Nicosia, University of Cyprus.

Rapoport, Y., and E. Savage-Smith, eds. 2014. *An Eleventh Century Egyptian Guide to the Universe, The Book of Curiosities*. Leiden and Boston: Brill. https://doi.org/10.1163/9789004256996

3. Naval Architecture.
The Hellenistic Hull Design
Origin and Evolution

Patrice Pomey

During the Hellenistic period and the Roman Republic, there was a dominant architectural design in Mediterranean naval architecture, which was eventually adopted by Greek and Roman shipyards as well as Punic. This system is characterised by a tripartite structure (keel, planking, framing). It allows the building of large ships with an elaborate hull shape, capable of good nautical performance, and this was clearly one of the factors behind the significant maritime expansion at the end of the first millennium BC. Of course, many other architectural approaches continued to be used in shipbuilding, which were testament to regional and local traditions.

The origin of this architectural system lies in the Greco-Roman evolution — between the second half of the sixth and the end of the fourth century BC — of sewn boats in the Greek tradition (influenced by the Phoenicians), which introduced the method of assembly by mortise and tenon. This resulted, throughout the Mediterranean, in a convergence of Greek and Phoenico-Punic architectural systems. However, during the Roman Empire, this approach — which presents a structural weakness at the level of the keel — was replaced by a new architectural type, apparently more robust, characterized by a flat bottom section, overlapping frames, and a keelson/mast step timber fitted on lateral sister-keelson.

> Nevertheless, the Hellenistic architectural system endured in the Eastern Mediterranean until at least the Early Byzantine period, as evidenced by the wreck of Yassiada 1.

In ancient naval architecture, the Hellenistic period is exceptional. It saw tremendous expansion in the field of shipbuilding, since the technical progress made over the previous centuries made it possible to build larger ships and more advanced hull shapes. Thus, the technical characteristics of the vessels were greatly improved in both terms of tonnage and sailing performance.[1]

Historical events helped to foster these developments and played a dynamic role in their evolution. The rivalries between the Hellenistic kingdoms, fuelled by the Alexandrian Empire, quickly led to an arms race, which resulted in the building of more and more large war galleys. From the trireme of the Classical period with its three rows of oars followed the quadrireme and the quinquereme, then the super-galleys of six and more. Appearing at the end of the fourth century BC in the fleet of Demetrios Poliorcetes, these super-galleys expanded quickly to reach the level of twenty and thirty rows of oars and culminate, towards the end of the third century BC, with the forty of Ptolemy IV Philopator powered by 4000 rowers.[2] Although the Romans were then confined to building quinqueremes, they built fleets of at least 100 and 200 units during the first Punic War.[3] Merchant vessels did not escape this phenomenon of gigantism and, in the third century BC, Hiero II of Syracuse, who wished to show the power of his country's shipyards and the richness of their local wheat fields, built the *Syracusia*, the largest grain ship of its time.[4] This was a ship purpose-built for trade but heavily armed, with three masts and three bridges, and with a crew of more than 825 people; its tonnage is estimated at 2000 to 4000 tonnes. Regardless of the method of calculation, this tonnage is huge and the launch of the ship required the assistance of Archimedes. But this technical exploit remained short-lived; the largest ships of the time in common use belong

---

1 Pomey 2011.
2 Casson 1971, 137–40; Basch 1987, 337–53; Pomey 2009.
3 Polybius, 1.20.13; 1.59.8.
4 Athenaeus, *Deipnosophistes* 5.206d-209b; for an analysis, see Casson 1971, 191–99; Pomey and Tchernia 2006; Pomey 2009; Nantet 2016, 126–31.

to the class of *myriophoroi*, carrying 10,000 amphorae, or 500 tonnes of deadweight.[5] These giant ships were exceptional, but their architecture remains largely unknown. Nonetheless, their existence reveals the technical capacity of large shipyards of the time.

However, thanks to a number of particularly representative wrecks, we can form an idea of the dominant architectural system in use in the Greek, Roman and Punic shipyards during the Hellenistic and Roman Republic. This system is based on a tripartite structure, used to construct ancient Mediterranean ships since at least the Archaic period. This structure is composed of an axial frame (keel, stem, sternpost), planking and transverse framing (frames, beams). Within this structure, the various elements can be given a number that define a particular architectural type, one we can call 'Hellenistic'. Among these features, some are common to all Hellenistic ships. They are said to be 'major' or 'primary'. Others, however, can be variable and are called 'secondary'.

The major characteristics are:

1. a cross-section with a *retour de galbord*, i.e. a cross-section with a wine-glass profile;

2. an axial frame, composed of a keel extended toward the extremities by the stem and sternpost; according to the ship's importance, end pieces can be more or less numerous and form stem and stern complexes;

3. a rabbeted keel and a carved polygonal garboard;

4. a carvel planking assembled by mortise-and-tenon joints;

5. a framing system composed of alternating floor timbers and half frames faced on the keel axis; floor timbers and half frames are extended by futtocks fitted with butt joint; all the frames elements are nailed or tree-nailed to the planking; the floor timbers, with some exceptions, are independent of the keel;

6. transverse beams supported by the wales of the planking;

7. an internal axial frame with a keelson/mast-step timber extended by a simple keelson; keelson and mast-step timber are fitted on the back of the floor timbers;

---

5   Pomey and Tchernia 1978.

8. a longitudinal internal framing composed of stringers nailed to the frames and mobile ceiling planks.

The secondary characteristics are:

1. a longitudinal section which may include an important rake; the bow shape (convex, straight or concave) varies according to the vessel type;

2. the stem complex may include a cutwater, and the stern complex a heel located under the sternpost in the extension of the keel, and acting like a drift board;

3. the planking might be single- or double-covered with lead sheathing.

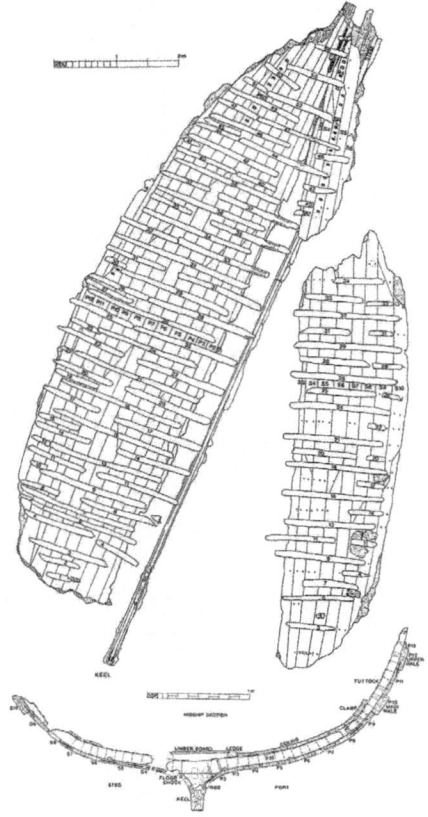

Figure 3.1 Kyrenia shipwreck. Plan and amidship cross-section (Steffy 1994).

3. Naval Architecture. The Hellenistic Hull Design

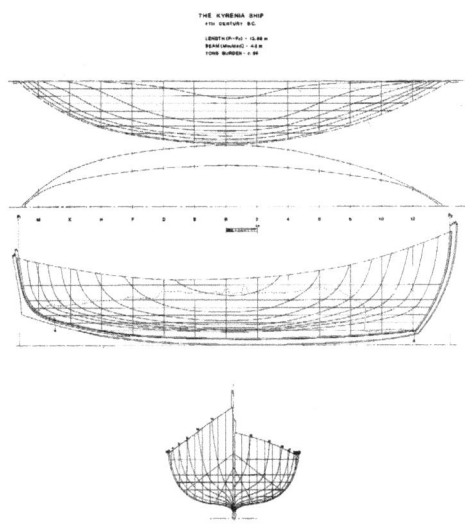

Figure 3.2 Kyrenia shipwreck. Reconstructed hull lines (Steffy 1994).

The first known example of this architectural type is the Kyrenia wreck, which is dated to the beginning of the third century BC, c. 295–285.[6] The shell, however, shows evidence of a number of repairs that indicate the boat enjoyed many seaworthy years before it sank. Therefore, it is possible to date its construction to c. 325–315 BC, and to hypothesise that it was therefore during this century that this architectural type was developed. The Kyrenia ship has the main features of the Hellenistic type, including the wine-glass cross-section, the planking assembled by tenon-and-mortise joints, alternate frames and the mast-step timber fitted on the back of the floor timbers (see figs. 3.1 and 3.2). It should be noted that the frames are fixed to the planking by means of clenched nails driven into wooden dowels, and lead sheathing has been set in afterwards to strengthen the hull and to complete its waterproofing. If it is not unique, the lead sheathing of the Kyrenia wreck is the first *known* example of this type of protection. In the following century, lead sheathing is mentioned on the *Syracusia* where it was set from the beginning of the construction, before the ship launched. It is also documented on the Punic wreck near Marsala (mid-third century BC) (Fig. 3.3).[7]

---

6   Wylde Swiny and Katzev 1973; Steffy 1985; 1994, 42–58; Womer Katzev 2005.
7   Frost 1976.

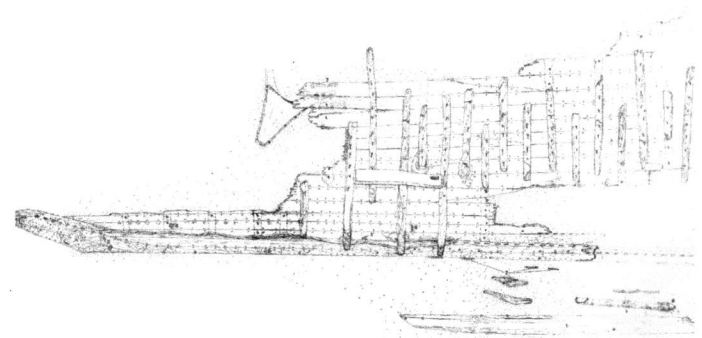

Figure 3.3 Marsala shipwreck. Hull plan (Frost 1976).

This last wreck has all the main features of the Hellenistic type, and it reflects the unification of architectural systems across the Mediterranean, from the Greco-Roman world to the Punic world. Nevertheless, within this architectural type, it is possible to find many variations depending on the secondary characteristics. The Roman ship dating from the first century BC (c. 75–60 BC), found wrecked in 1967 off Madrague de Giens on the Giens peninsula, illustrates the high degree of sophistication attained by this type (of which it is likely to be one of the best examples) (Fig. 3.4).[8]

Figure 3.4 Madrague de Giens shipwreck. General view of hull (Photo A. Chéné, AMU, CNRS, MCC, CCJ).

---

8   Tchernia et al. 1978, 75–99; Pomey 1982; Liou and Pomey 1985, 553–56; Pomey 2004b, 2015.

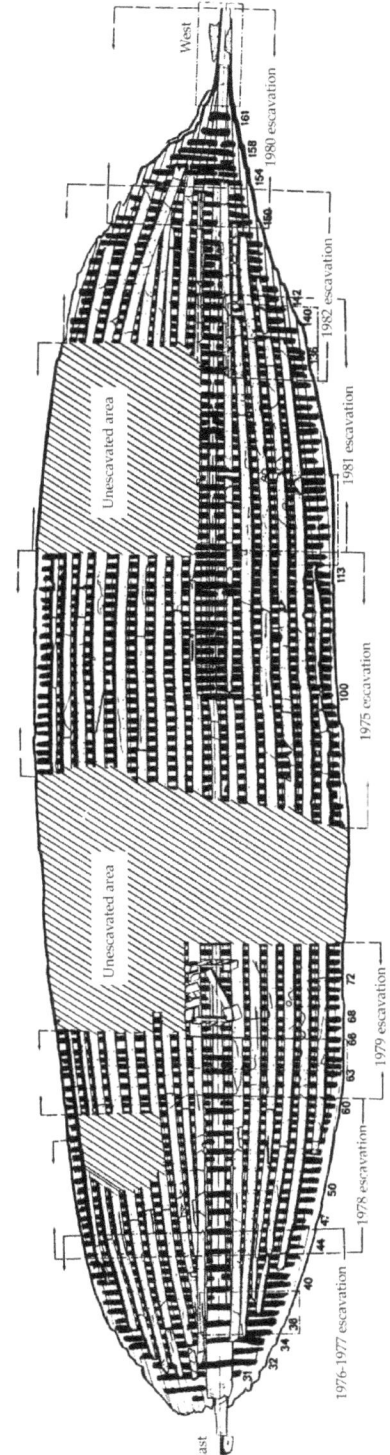

Figure 3.5 Madrague de Giens shipwreck. Plan of the hull remains (Drawing M. Rival, AMU, CNRS, MCC, CCJ).

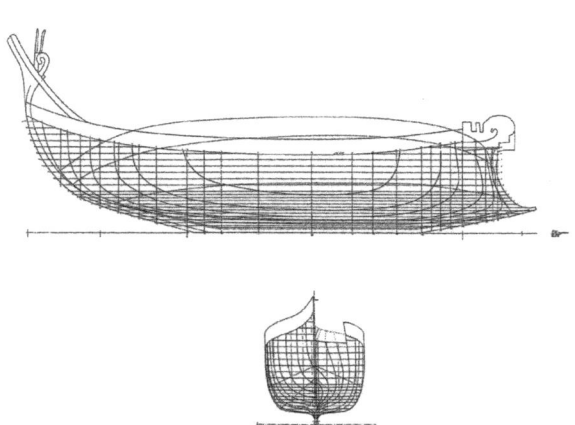

Figure 3.6 Madrague de Giens shipwreck. Reconstructed hull lines (Drawing M. Rival, AMU, CNRS, MCC, CCJ).

According to its dimensions, nearly 40 m long, 9 m wide and 4.50 m deep, and its tonnage, estimated to be 400 tonnes deadweight,[9] the ship belongs to the category of *myriophoroi* (Figs. 3.5 and 3.6).[10]

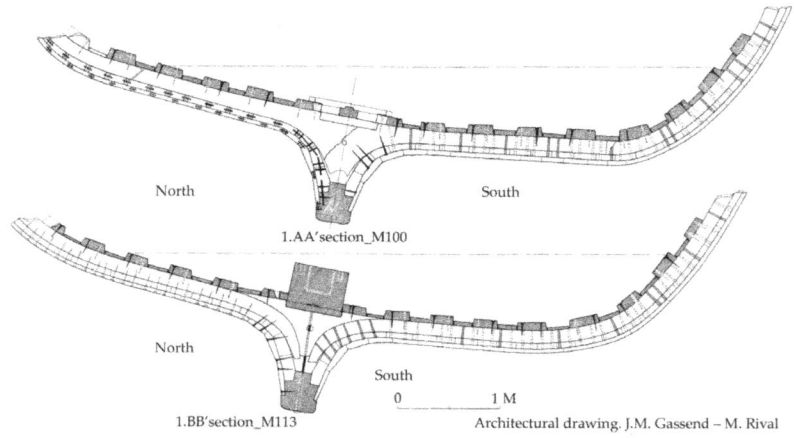

Figure 3.7 Madrague de Giens shipwreck. Amidship cross-sections (Drawing J.-M. Gassend, M. Rival, AMU, CNRS, MCC, CCJ).

---

9   Nantet 2016, 355–60.
10 Pomey and Tchernia 1978.

3. *Naval Architecture. The Hellenistic Hull Design*   35

Its structure, characteristic of the Hellenistic type, is distinguished by its elaborate forms, its stem and stern complex and its double planking reinforced by lead sheathing (figs. 3.7, 3.8, 3.9 and 3.10).

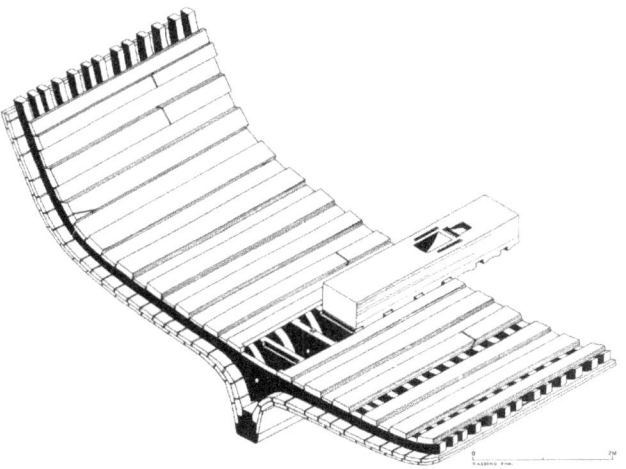

Figure 3.8 Madrague de Giens shipwreck. General axonometric view (Drawing J.-M. Gassend, M. Rival, AMU. CNRS, MCC, CCJ).

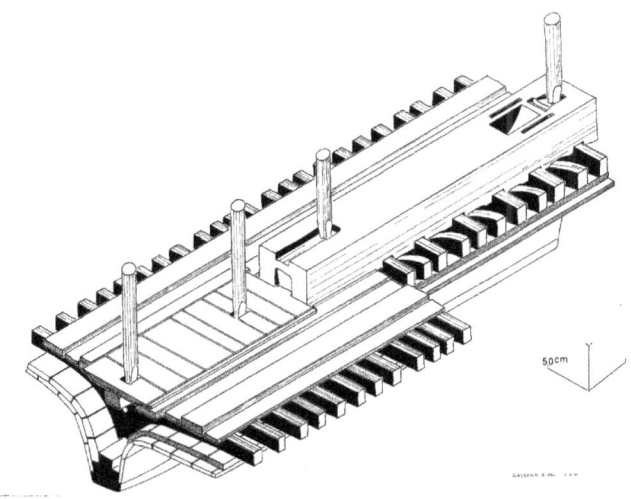

Figure 3.9 Madrague de Giens shipwreck. Axial axonometric view (Drawing J.-M. Gassend, M. Rival, AMU, CNRS, MCC, CCJ).

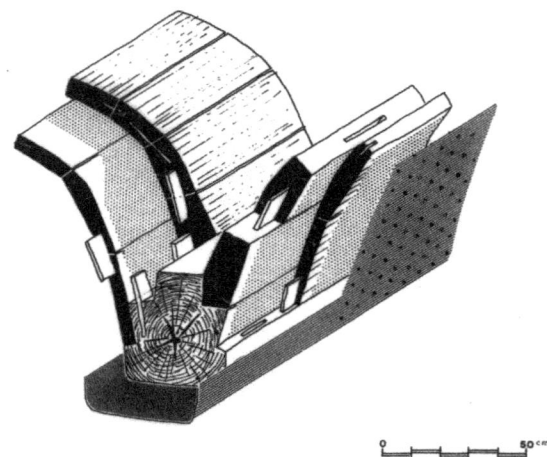

Figure 3.10 Madrague de Giens shipwreck. Axonometric view of the keel, the double planking and the hull sheathing (Drawing J.-M. Gassend, M. Rival, AMU, CNRS, MCC, CCJ).

The stern complex includes no fewer than six frame pieces, buttressed one by the other, in order to give structural rigidity to the long rake aft. Among these pieces a stern heel, located under the sternpost and in the extension of the keel, acts as a drift spoiler (Fig. 3.11).

As for the stem complex, located in the extension of a long raised forefoot, it has an inverted (convex) stem extended on the front by a cutwater. The whole, formed by the prominent keel, the drift spoiler and the cutwater, forms a very important drift plan, which was to make the ship very stable at all sailing trims (Fig. 3.12). Finally, it should be noted that a number of floor timbers are attached to the keel by a strong copper bolt (Fig. 3.13). This is the oldest known example of the use of such bolts that has been discovered, although they are described as being used in the earlier *Syracusia*. However, the floor timbers of the Madrague de Giens wreck, including the bolted ones, do not touch the keel, and remain largely independent. The few bolts therefore appear to be reinforcing the keel/floor-timber link to remedy the structural weakness of the longitudinal axis, due to the prominence of the keel and the independence of the floor timbers.[11] In fact, the examination of

---

11 Because of the many repairs that are evident on the hull bottom of the Madrague de Giens ship, and the traces that can be observed on the floor timbers, it is very possible that the keel was changed, no doubt after some kind of physical shock that affected it.

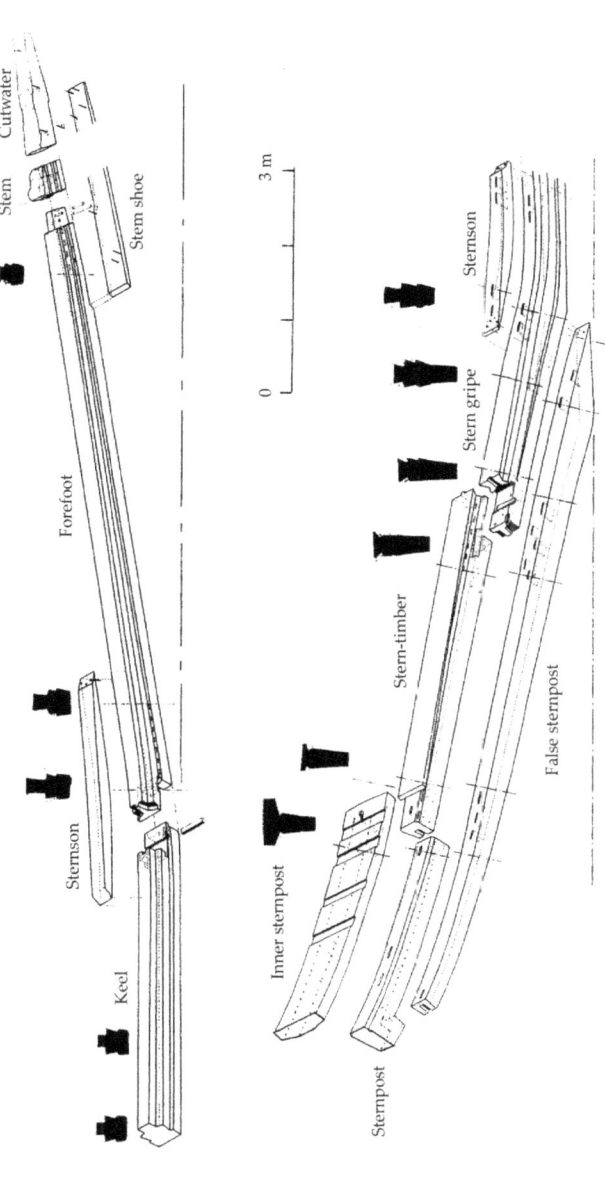

Figure 3.11 Madrague de Giens shipwreck. Axonometric views of the stem complex (top) and the stern complex (bottom) (Drawing M. Rival, AMU, CNRS, MCC, CCJ).

these bolted floor timbers shows that they were not pre-erected, and so they do not call into question the longitudinal conception and the shell construction principle of the ship.[12]

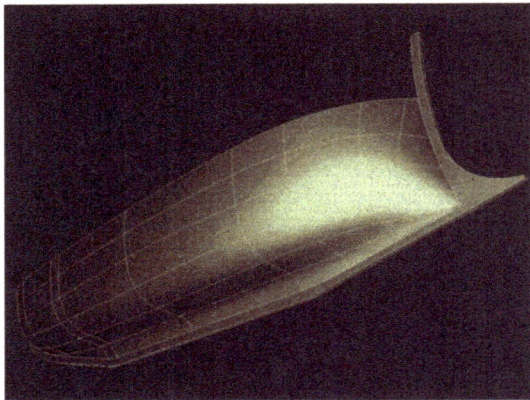

Figure 3.12 Madrague de Giens shipwreck. 3D reconstruction of the hull shapes (Drawing Sistre international).

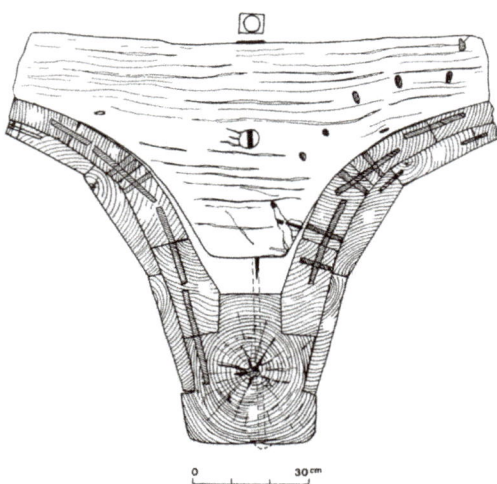

Figure 3.13 Madrague de Giens shipwreck. Detailed section of the keel area. Note the bolt joining the floor-timber to the keel (Drawing J.-M. Gassend, M. Rival, AMU, CNRS, MCC, CCJ).

12   Pomey 1988, 406; 1998, 66; 2004a, 30–31.

As to the construction process, everything indicates that the ship was probably built using a 'shell first' method.

If questions arise about the wreck of Madrague de Giens because of the presence of bolted floor timbers, it is clear that the Kyrenia wreck was entirely conceived and realized 'shell first', as was the Marsala wreck.[13] As a result, everything indicates that the Hellenistic type of vessel was conceived according to the principle of the longitudinal and 'shell' construction and built according to the 'shell first' process.

As we can see, this dominant architectural system allowed for the building of ships of large size with an elaborate hull shape, capable of good nautical performance. The system also seems well adapted to the construction of coasters (Kyrenia) as well as oceangoing vessels (Madrague de Giens) or warships (Marsala). It was adopted for use in both private and state shipyards. In addition to the wrecks previously discussed, the Hellenistic type can be identified with its variants on the following wrecks: Baie de Briande (first half of the second century BC), Grand Congloué (second century BC), Caveaux I (end of the second — beginning of the first century BC), Cavalière (c. 100 BC), Mahdia (beginning of the first century BC), Albenga (first half of the first century BC), Pointe de Pomègues (first half of the first century BC), Chrétienne A (c. 75 BC), Dramont A (mid-first century BC), Titan (mid-first century BC), Plane I (mid-first century BC) (Fig. 3.14).

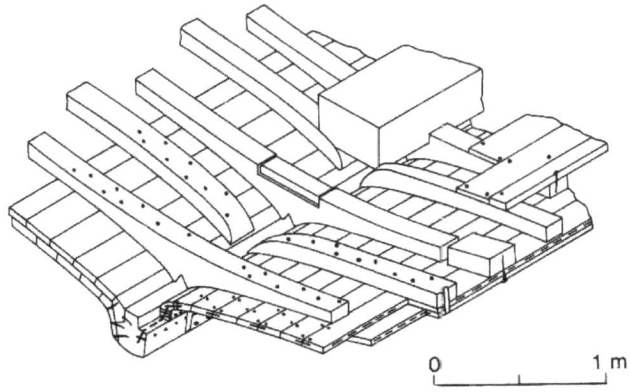

Figure 3.14 Dramont A shipwreck. Axonometric view of the central part of the hull (Drawing Cl. Santamaria).

---

13   Pomey 1988, 405–06; 1998; 2004a, 29–30.

This architectural system is clearly one of the factors that led to the significant maritime expansion of the end of the first millennium BC. Of course, this type, although dominant, did not preclude the existence of many other architectural types that attest to regional and local traditions.

The origin of this architectural system lies in the Greco-Roman evolution — between the second half of the sixth and the end of the fourth century BC — of sewn boats in the Greek tradition.[14] According to the most ancient archaeological examples we have — including the shipwrecks of Giglio, Pabuc Burnu, Cala Sant Vicenç, Bon-Porté 1 and Jules-Verne 9, all dating back to the sixth century BC[15] — ancient Greek ships were entirely assembled by ligatures (figs. 3.15, 3.16 and 3.17).

Figure 3.15 Jules-Verne 9 shipwreck. General view of the hull remains (Photo M. Derain, AMU, CNRS, MCC, CCJ).

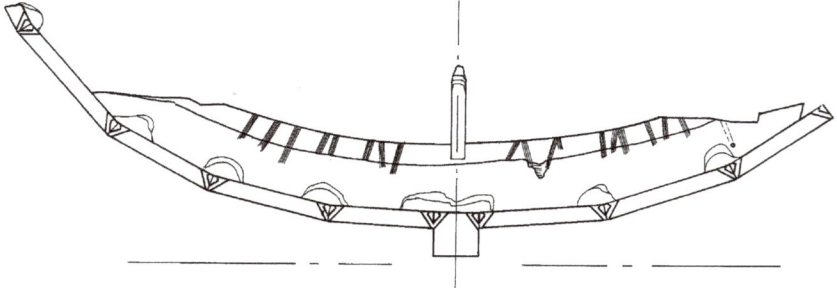

Figure 3.16 Jules-Verne 9 shipwreck. Cross-section of the hull remains (Drawing M. Rival, AMU, CNRS, MCC, CCJ).

---

14 Pomey 1997; Kahanov and Pomey 2004; Pomey 2010.
15 Although the literary testimonies of Homer suggest that this tradition could go back to the Bronze Age; see *Iliad* 135; *Odyssey* 5.244–57.

3. Naval Architecture. The Hellenistic Hull Design   41

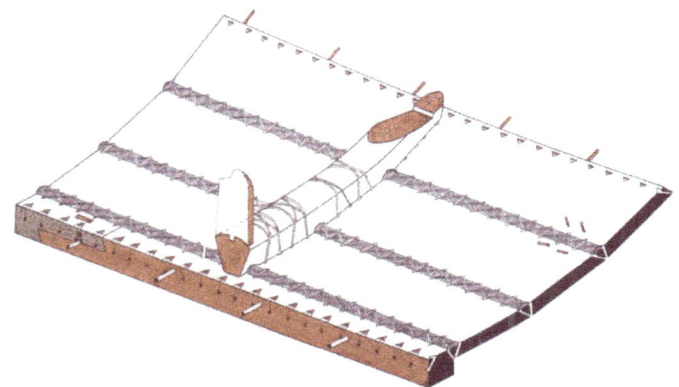

Figure 3.17 Jules-Verne 9 shipwreck. Axonometric view of the sewing and lashing of the hull assembly system (Drawing M. Rival, AMU, CNRS,MCC, CCJ).

During the first transitionary phase, illustrated by the shipwrecks Jules-Verne 7, Villeneuve-Bargemon 1 (or Caesar 1), Grand Ribaud F and Gela 1, the assembly systems by tenon-and-mortise joint for the planking and by nailing for the frames emerges (Figs. 3.18, 3.19, 3.20, 3.21 and 3.22).

Figure 3.18 Jules-Verne 7 shipwreck. General view of the hull remains (Photo M. Derain, AMU, CNRS, MCC, CCJ).

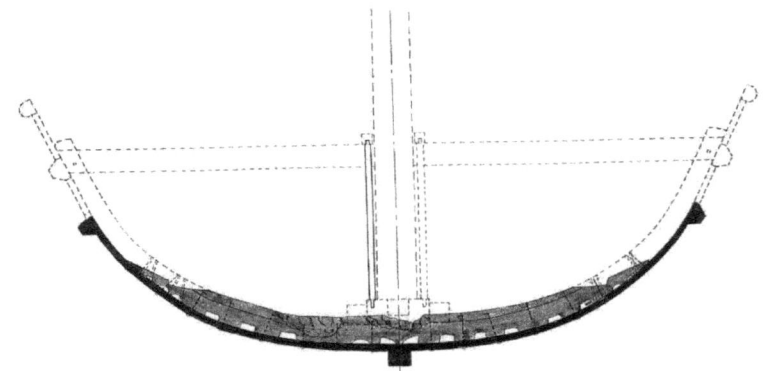

Figure 3.19 Jules-Verne 7 shipwreck. Amidship cross-section of the hull remains (Drawing M. Rival, AMU, CNRS, MCC, CCJ).

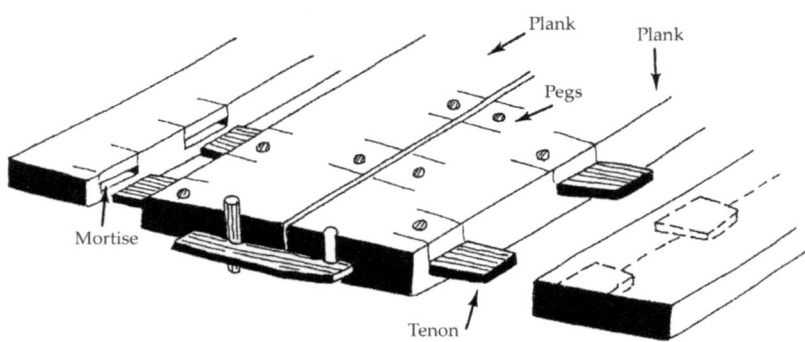

Figure 3.20 Theoretical schema of the mortise-and-tenon joint (Drawing M. Rival, AMU, CNRS, MCC, CCJ).

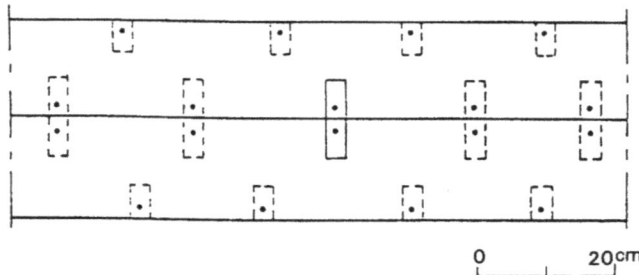

Figure 3.21 Jules-Verne 7 shipwreck. Schema of the mortise-and-tenon joint network (Drawing M. Rival, AMU, CNRS, MCC, CCJ).

3. Naval Architecture. The Hellenistic Hull Design   43

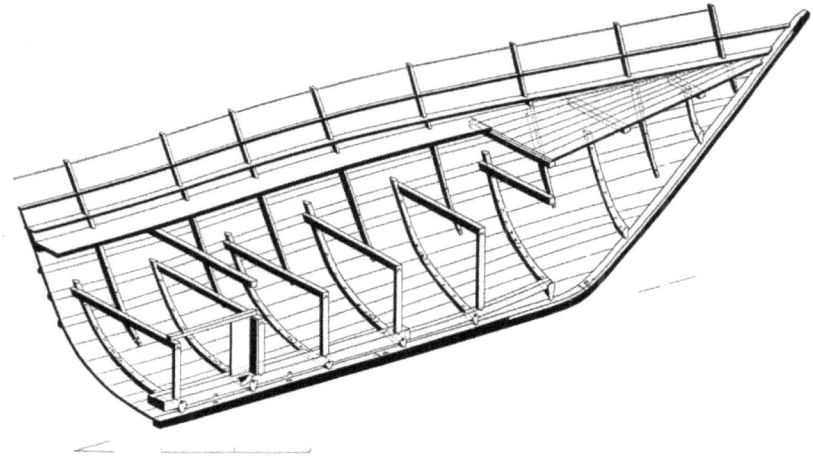

Figure 3.22 Jules-Verne 7 shipwreck. General axonometric view of the hull structure (towards the bow). Note the framing with floor-timbers alternating with top timbers; the mast step timber fitted on the floor-timbers; the beams fitted on the extremities of the floor-timbers (Drawing M. Rival, AMU, CNRS, MCC, CCJ).

They come as a substitute for the previous sewing and ligatures. However, the sewing did not disappear completely and was still used for some parts of the ship, mainly at first for the extremities, and then for repairs. In the second phase of the development of the Hellenistic type, defined by the shipwrecks Gela 2 and Ma'agan Mikhael, the use of sewing becomes even less common in favour of the development of the tenon-and-mortise joint. The hull shapes begin to evolve and the hull bottom, previously round, starts to present a wine-glass cross-section (Figs. 3.23, 3.24 and 3.25). Finally, the last stage of development is provided by the Kyrenia wreck: the seams and ligatures have totally disappeared, except in some reused planks; the hull cross-section is now a wine-glass shape and the keel is completely rabbeted. The frames, originally trapezoidal in order to be strongly lashed, are rectangular and nailed to the planking; while the top timbers, located between the floor timbers and formerly implanted only in the top of the wall, are extended to the hull bottom in order to form half frames faced on the keel axis.

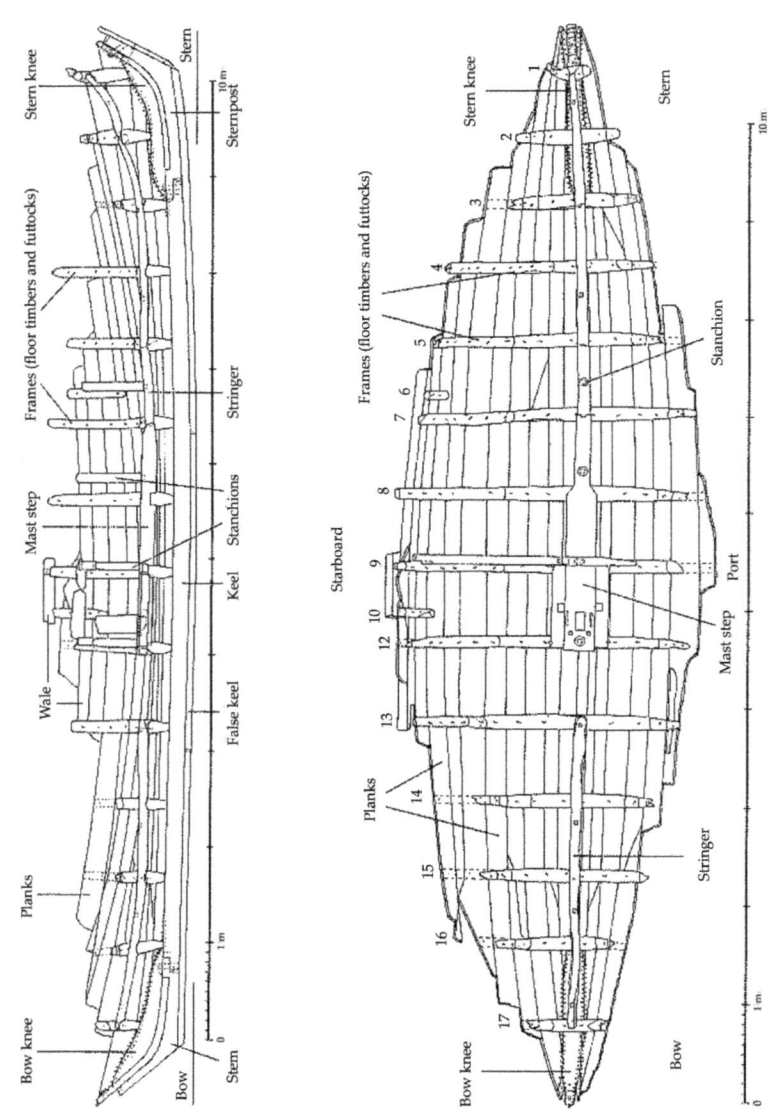

Figure 3.23 Ma'agan Mikhael shipwreck. Plan and longitudinal section of the hull remains (Kahanov, Linder 2004).

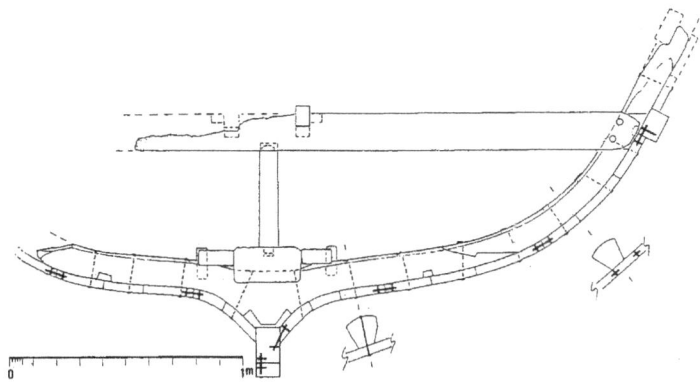

Figure 3.24 Ma'agan Mikhael shipwreck. Main cross-section of the hull remains (From Kahanov, Linder 2004).

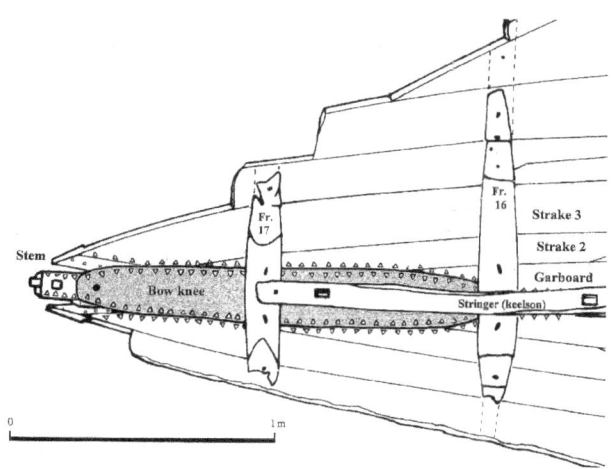

Figure 3.25 Ma'agan Mikhael shipwreck. Top view of the bow with the sewn bow knee (Kahanov, Linder 2004).

The mast-step timber, since the earliest evidence of Greek sewn boats of the sixth century BCE and according to the example provided by the Bon-Porté wreck, was directly fitted on the back of the floor timbers. Thus, the Kyrenia wreck, which is located at the end of the chain of evolution of Greek shipbuilding, already has the main characteristics that define the Hellenistic type, and represents the prototype.

It is obvious that the replacement of the sewing of the planking by tenon-and-mortise joints and of the ligatures of the framing by nails or treenails contributed significantly to enhance the strength of the hulls and their longevity. This evolution led to the building of larger ships with a greater tonnage, and with significantly evolved hull shapes, which enabled the development of new ship types like the trireme (Fig. 3.26).[16] As to the origin of this evolution, which led to the introduction of the tenon-and-mortise joint in the Greek tradition, it appears more likely that it is a Punic influence, where this system had been in use since the Bronze Age,[17] rather than an internal evolution.[18] That explains the convergence across the entire Mediterranean of Greek and Phoenico-Punic traditions, leading to the Hellenistic type that was used in the Greco-Roman and Punic worlds.

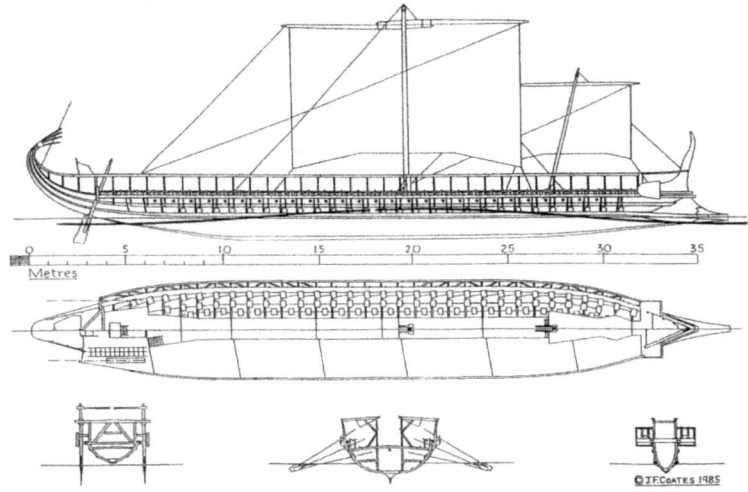

Figure 3.26 Trireme replica *Olympias*. General plans (J.F. Coates).

However, if the Hellenistic type offers some undeniable sailing qualities, as seen through the Madrague de Giens vessel, it presents nevertheless a structural weakness at the keel level. This weakness is due to the prominence of the keel, characteristic of the wine-glass cross-section, and to the lack of connection between the keel and the floor timbers. Many shipwrecks (Pointe de Pomègues, Plane I, Caveaux I, Baie de Briande, Chrétienne A, to name only those found off French coasts), which sank after losing their keel following a shock, testify eloquently to this problem (Fig. 3.27).

---

16   Pomey 2011.
17   Pomey 1997; Kahanov and Pomey 2004; Pomey 2010; Pomey and Boetto forthcoming.
18   Polzer 2010, 2011.

3. Naval Architecture. The Hellenistic Hull Design 47

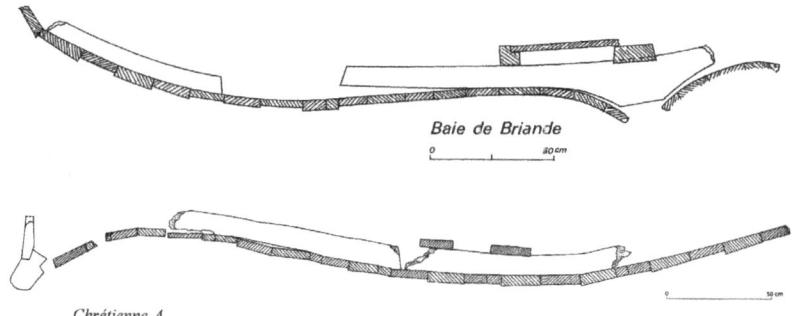

Figure 3.27 a- Baie de Briande shipwreck; b- Chrétienne A shipwreck. Note the rupture or the loss of the keel (Drawing M. Rival, AMU, CNRS, MCC, CCJ).

We have seen, in the case of the Madrague de Giens wreck, that this is most likely what led to the development of the practice of bolting the floor timbers.[19] This practice probably prefigured the use of pre-built active frames that marked the beginning of the evolution towards a skeleton construction.[20] But it is perhaps also the reason for the advent under the Roman Empire of a new architectural type of ship, with flat floor timbers, which mainly originated in the Western Mediterranean, and for this reason was called the Western Imperial Roman type.[21] With a relatively flat bottom without prominent keel, internal framing strengthened with numerous floor timbers bolted to the keel, overlapping half frames and a mast-step timber fitted on two sister keelson, this ship type, with a large loading capacity, should have been structurally stronger, although not necessarily better in terms of its qualities for sailing (Fig. 3.28). It is probably due to these nautical qualities that the Hellenistic type did not become extinct within the time of the Mediterranean trade fleets, surviving in the Eastern Mediterranean until the Byzantine period, as shown in the wrecks of Yassiada 2 (fourth century AD),[22] Yassiada 1 (seventh century AD)[23] and Bozburun (ninth century AD)[24] (Figs. 3.29, 3.30 and 3.31).

---

19 Pomey 2002; Pomey 2011, 53–55.
20 Pomey et al. 2012.
21 Pomey 1998, 68; Pomey and Rieth 2005, 165–67; Pomey et al. 2012, 298–303.
22 van Doorninck 1976.
23 Bass and van Doorninck, 1982.
24 Harpster 2009.

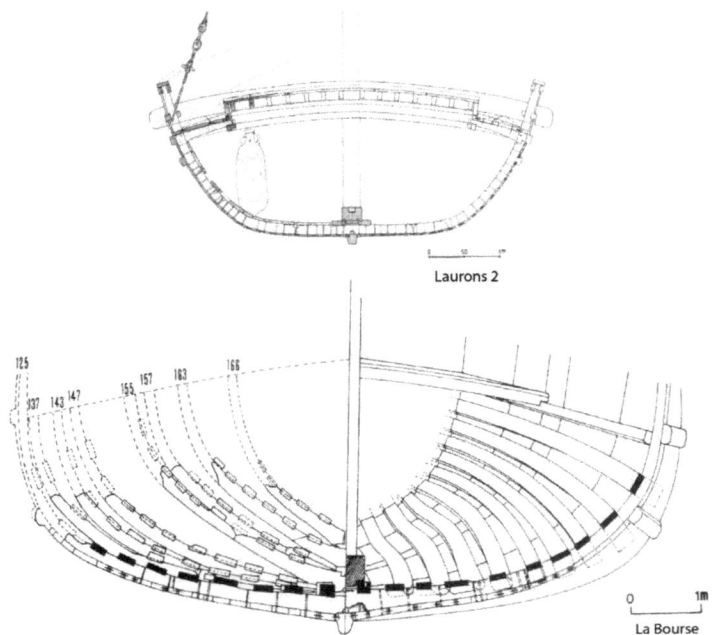

Figure 3.28 Western Roman Imperial type: top- Laurons 2 shipwreck; bottom- La Bourse shipwreck (Marseilles) (P. Pomey, AMU, CNRS, MCC, CCJ).

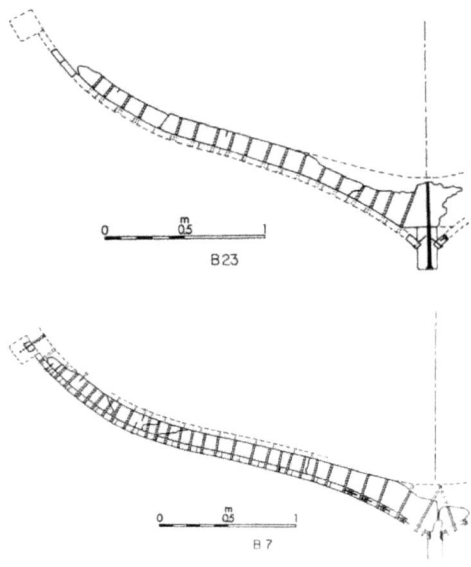

Figure 3.29 Yassiada 2 shipwreck. Cross-sections at frame B7 and B23 (van Doorninck 1976).

3. Naval Architecture. The Hellenistic Hull Design   49

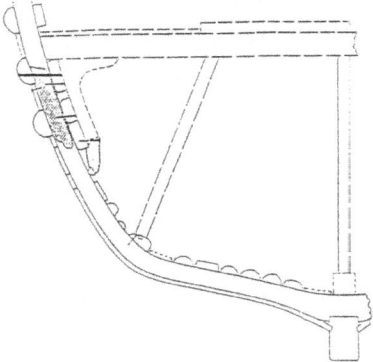

Figure 3.30 Yassiada 1 shipwreck. Amidship cross-sections (Steffy 1982).

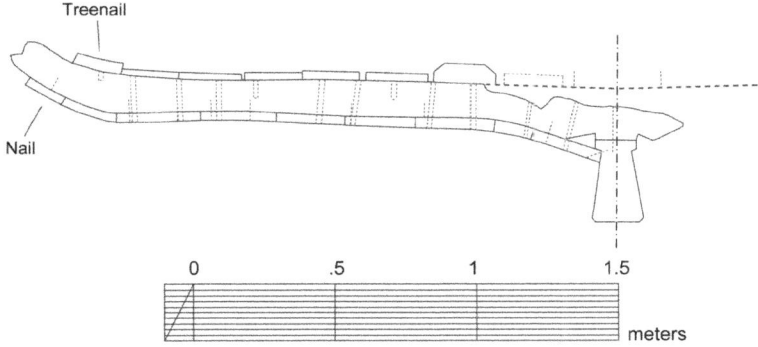

Figure 3.31 Bozborum shipwreck. Cross-section of the hull (floor-timber 1) (Harpster 2002).

Similarly, in Roman nautical iconography, including in Africa, the Hellenistic type is always represented, as can be seen on the mosaic of the *Syllectani* in the *Piazzale delle Corporazioni* in Ostia Antica (end of the second century AD), and on the mosaic of the *frigidarium* of the baths of Themetra (Tunisia, third century AD), whose large vessels represent similar ships to the Madrague de Giens[25] (Figs. 3.32, 3.33 and 3.34).

---

25   Pomey 1982.

Figure 3.32 Mosaic of the *frigidarium* of the bath of Themetra (Tunisia, 3rd c. AD). Ship of Madrague de Giens type (Photo R. Guéry, AMU, CNRS, MCC, CCJ).

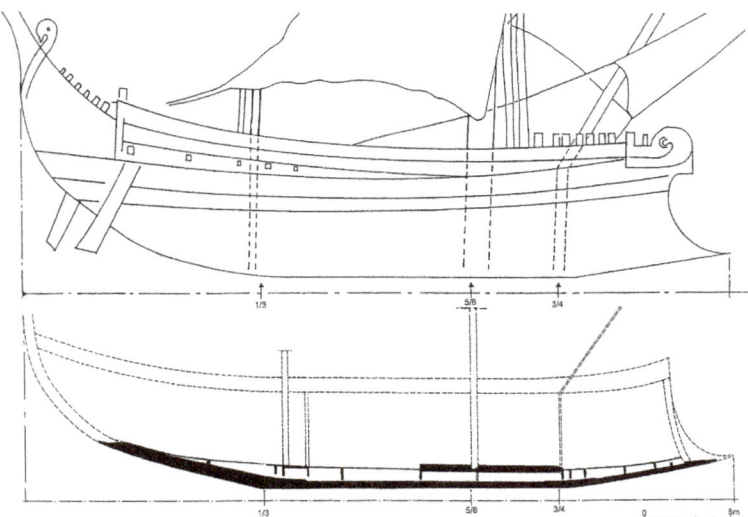

Figure 3.33 Comparative sketch of the Themetra ship (top) and the Madrague de Giens (below). Note the similarity of the hull profiles (Drawing M. Rival, AMU, CNRS, MCC, CCJ).

Figure 3.34 Mosaic of the *Syllectani* in the *Piazzale delle Corporazioni* (Ostia Antica, late 2nd c. AD). Note the similarity of profile between the ship on the left and the Madrague de Giens (Photo A. Chéné, AMU, CNRS, MCC, CCJ).

Bibliography

Basch, L. 1987. *Le musée imaginaire de la marine antique*. Athens: Institut hellénique pour la préservation de la tradition nautique.

Bass, G. F., and F. H. Jr. van Doorninck, eds. 1982. *Yassi Ada Volume 1. A Seventh-Century Byzantine Shipwreck*. College Station: Texas A&M University Press.

Casson, L. 1971. *Ships and Seamanship in the Ancient World*. 2nd ed. Princeton: Princeton University Press.

Frost, H. 1976. *Lilybaeum (Marsala). The Punic Ship: Final Excavation Report. Notizie degli Scavi di Antiquità* Suppl. 30. Rome: Accademia nazionale dei Lincei.

Harpster, M. 2009. 'Designing the 9th-Century-AD Vessel from Bozburun.' *International Journal of Nautical Archaeology and Underwater Exploration* 38: 297–313. https://doi.org/10.1111/j.1095-9270.2009.00226.x

Kahanov, Y., and E. Linder. 2004. *The Ma'agan Mikhael Ship. The Recovery of a 2400-Years-Old Merchantman*. Vol. 2. *Final Report*. Haifa: Israel Exploration Society.

Kahanov, Y., and P., Pomey. 2004. 'The Greek Sewn Shipbuilding Tradition and the Ma'agan Mikhael Ship: A Comparison with Mediterranean Parallels from the Sixth to the Fourth Centuries BC.' *The Mariner's Mirror* 90: 6–28. https://doi.org/10.1080/00253359.2004.10656882

Nantet, E. 2016. *Phortia. Le Tonnage des navires de commerce en Méditerranée du VIIIe siècle av. l'ère chrétienne au VIe siècle de l'ère chrétienne*. Rennes: Presses Universitaires de Rennes.

Liou, B., and P. Pomey. 1985. 'Informations archéologiques: recherches sous-marines.' *Gallia* 43: 547–76.

Polzer, M. E. 2010. 'The VIth-Century B.C. Shipwreck at Pabuç Burnu, Turkey: Evidence for Transition from Lacing to Mortise-and-Tenon Joinery in Late Archaic Greek Shipbuilding.' In *Transferts technologiques en architecture navale méditerranéenne de l'Antiquité aux temps modernes: identité technique et identité culturelle, Actes de la Table Ronde, Istanbul 2007*, edited by P. Pomey: 27–44. Varia Anatolica 20. Istanbul: Institut français d'études anatoliennes-Georges-Dumézil.

Polzer, M. E. 2011. 'Early Shipbuilding in the Eastern Mediterranean.' In *The Oxford Handbook of Maritime Archaeology*, edited by A. Catsambis, B. Ford, and D. Hamilton, 349–78. New York: Oxford University Press. https://doi.org/10.1093/oxfordhb/9780199336005.013.0016

Pomey, P. 1982. 'Le navire romain de la Madrague de Giens.' *Comptes-Rendus de l'Académie des Inscriptions et Belles: CRAI* 126(1): 133–54.

Pomey, P. 1988. 'Principes et méthodes de construction en architecture navale antique.' In *Navires et commerces de la Méditerranée antique, Hommage à Jean Rougé. Cahiers d'Histoire* 33: 397–412.

Pomey, P. 1998. 'Conception et réalisation des navires dans l'Antiquité méditerranéenne.' In *Concevoir et construire les navires. De la trière au picoteux*, edited by E. Rieth, 49–72. Technologie, Idéologies, Pratique, Revue d'anthropologie des connaissances. Ramonville-Sainte-Agne: Erès.

Pomey, P. 1997. 'Un exemple d'évolution des techniques de construction navale antique: de l'assemblage par ligatures à l'assemblage par tenons et mortaises.' In *Techniques et économie antiques et médiévales, 'Le temps de l'innovation', Colloque international, Aix-en-Provence 1996*, edited by D. Garcia and D. Meeks, 195–203. Paris: Errance.

Pomey, P. 2002. 'Remarque sur la faiblesse des quilles des navires antiques à retour de galbord.' In *Vivre, produire et échanger: reflets méditerranéens. Mélanges offerts à B. Liou*, edited by L. Rivet and M. Sciallano, 11–19. Montagnac: Monique Mergoil.

Pomey, P. 2004a. 'Principles and Methods of Construction in Ancient Naval Architecture.' In *The Philosophy of Shipbuilding. Conceptual Approaches to the Study of Wooden Ships*, edited by F. M. Hocker and C. A. Ward, 25–36. College Station: Texas A&M University Press.

Pomey, P. 2004b. 'La structure du navire de la Madrague de Giens et le type hellénistique.' *Ligures, Rivista di Archeologia, Storia, Arte e Cultura Ligure* 2: 370–73.

Pomey, P. 2009. 'Sur les eaux d'Alexandrie: des navires et des bateaux.' In *Du Nil à Alexandrie. Histoire d'eaux*, edited by I. Fairy, 514–35. Alexandria: Harpocrates.

Pomey, P. 2011. 'Les conséquences de l'évolution des techniques de construction navale sur l'économie maritime: quelques exemples.' In *Maritime Technology in the Ancient Economy: Ship-Design and Navigation*, edited by W. V. Harris and K. Iara, 39–55. *Journal of Roman Archaeology* Suppl. 84.

Pomey, P. 2015. 'The Madrague de Giens Project in the Wake of the Excavation of the Byzantine Shipwreck at Yassiada.' In *Maritime Studies in the Wake of the Byzantine Shipwreck at Yassiada, Turkey*, edited by D. N. Carlson, J. Leidwanger, S. M. Kampbell, 73–81. College Station: Texas A&M University Press.

Pomey, P., and G. Boetto, forthcoming. 'Ancient Mediterranean Sewn Boats Traditions.' In *Fibre and Wood. Sewn Boat Construction Techniques through Times, Muscat, February 2015*, edited by L. Blue and E. Staples.

Pomey, P., Y. Kahanov, and E. Rieth. 2012. 'Transition from Shell to Skeleton in Ancient Mediterranean Ship-Construction: Analysis, Problems, and Future Research.' *International Journal of Nautical Archaeology and Underwater Exploration* 41: 235–314. https://doi.org/10.1111/j.1095-9270.2012.00357.x

Pomey, P., and E. Rieth. 2005. *L'archéologie navale*. Paris: Errance.

Pomey, P., and A. Tchernia. 1978. 'Le tonnage maximum des navires de commerce romains.' *Archaeonautica* 2: 233–51.

Pomey, P., and A. Tchernia. 2006. 'Les inventions entre l'anonymat et l'exploit: le pressoir à vis et la Syracusia.' In *Innovazione tecnica e progresso economico nel mondo romano: atti degli Incontri capresi di storia dell'economia antica (Capri, 13–16 aprile 2003)*, edited by E. Lo Cascio, 81–99. Bari: Edipuglia.

Steffy, J. R. 1985. 'The Kyrenia Ship: An Interim Report on its Hull Construction.' *American Journal of Archaeology* 89: 71–101.

Steffy, J. R. 1994. *Wooden Ship Building and the Interpretation of Shipwrecks*. College Station: Texas A&M University Press.

Tchernia, A., P. Pomey, A. Hesnard, et al. 1978. *L'Épave romaine de la Madrague de Giens. Gallia* Suppl. 34. Paris: Éditions du Centre national de la recherche scientifique.

van Doorninck, F. H. Jr. 1976. 'The 4th Century Wreck at Yassi Ada. An Interim Report on the Hull.' *International Journal of Nautical Archaeology and Underwater Exploration* 5: 115–31.

Womer Katzev, S. 2005. 'Resurrecting an Ancient Greek Ship: Kyrenia, Cyprus.' In *Beneath the Seven Seas: Adventures with the Institute of Nautical Archaeology*, edited by G. F. Bass, 72–9. New York: Thames & Hudson.

Wylde Swiny, H., and M. L. Katzev. 1973. 'The Kyrenia Shipwreck: A Fourth-Century B.C. Greek Merchant Ship.' In *Marine Archaeology*, edited by D. J. Blackman, 339–55. *Colston Papers* 23. London: Butterworths.

4. Naves Pingere

'Painting Ships' in the Hellenistic Period

Martin Galinier and Emmanuel Nantet

Ancient literary sources often mention the existence of 'ship painters'. What did this expression mean exactly? Were these artists representing ships in their paintings, or were they craftsmen who were adorning ships? The reappraisal of these texts gives us the opportunity to consider the two different situations. Indeed, during the Hellenistic period there were a great deal of marine paintings displaying ships. Alongside these famous painters, the numerous craftsmen who were devoted to the adornment of ships remained anonymous. Only a very few of them could overcome the stigma attached to the label of 'craftsman' and produce paintings too: one such painter was Protogenes.

'Painting ships' (*naves pingere*), as Pliny the Elder said about the activities of the painter Protogenes ('until the age of fifty he was also a ship painter…'),[1] may have two meanings: the first is to adorn ships; the second consists of representing ships in paintings. During Alexander's funerals, the painter Apelles — or his workshop — may have produced four panels adorning the hearse of the conqueror and celebrating the military power of the Macedonian. Among these depictions was his war fleet.[2] However, this picture, linked to the event that defines the early

---

1 Pliny the Elder, *Natural History* 35.101.
2 Diodorus Siculus, 18.27: 'the fourth, ships made ready for naval combat.' In addition to a war fleet, a skit represented Alexander holding a sceptre surrounded by Macedonians and Persians; another with elephants mounted by Indians and Macedonians; and the third with horsemen.

Hellenistic period, is neither the first nor the only one to represent ships. This chapter will explore these two very different pictorial exercises over a long period of time.

Few artists are described as 'ship painters'. As a prefect of the imperial war fleet in Misenum, Pliny knew ships very well. Therefore, in calling someone a ship painter, he might imply both artistic and technical expertise. We should therefore pay particular attention to the work of Protogenes. What precisely was a ship painter in the ancient Mediterranean?[3]

4.1 Naval Issues Before the Reign of Alexander

The Ancient Greeks, as far as we can glean from textual evidence, had a close interest in the sea and its navigation. If *Iliad* includes 'The Catalogue of Ships',[4] *Odyssey* often describes the sea as barren and bitter. When Ulysses is about to leave Circe, a long description is dedicated to the construction of his raft, to the choice of the timbers, to the techniques used (with Circe's help).[5] Likewise, when Ulysses arrives among the Phaeacians, he notices the harbour and the ships in this city to which Poseidon granted 'the great gulf of the sea (...)'.[6] The launch of the ship offered to Ulysses by the king of the Phaeacians, Alkinoos, is also accurately described using technical details.[7]

In the literary sources describing easel paintings, major works most of which have not survived to the present day, several references to maritime and naval representations can be found. Achilles' shield, in *Iliad* depicts the god Ocean as the border of the world,[8] as does Herakles' shield in Hesiod (it also includes a 'harbour with a safe haven'[9]). The

---

3   See primarily Reinach 1921. More recently, Rouveret 2017, 61–84.
4   Homer, *Iliad* 2.
5   Homer, *Odyssey* 5.160–269. Pamphilus of Amphipolis (400–350) represented Odysseus on his skiff (Pliny the Elder, *Natural History* 35.76) On Odysseus' craft, see Casson 1964, 61–64; Casson 1992, 73–74; Mark 1991, 441–45; Mark 1996, 46–48; Mark 2005, 70–96; Tchernia 2001, 625–31.
6   Homer, *Odyssey* 7.35.
7   Homer, *Odyssey* 8.50 and following verses: '[...] they drew the black ship down to the deep water, and placed the mast and sail in the black ship, and fitted the oars in the leathern thole-straps, all in due order, and spread the white sail. Well out in the roadstead they moored the ship [...]'.
8   Homer, *Iliad* 16.
9   Hesiodos, *The Shield of Heracles* 207–08.

'Catalogue of Ships' enumerates all the black ships of the Achaeans taking part in the Trojan War.

Some artistic works are naturally inspired by the Homeric corpus. Indeed, the episode of the 'Battle of the Greeks and the Trojans close to the ships' is, according to Pausanias, described on the Chest of Kypselos,[10] an *ex-voto* carried out to the temple of Hera in Olympia in the sixth century BCE. However, one of the most ancient literary references to paintings can be found in Herodotus. It is noticeable that this reference consists of an historical anecdote: the painting (*graphsamenos*) is a present offered by Darius I to Mandrokles of Samos, in order to reward him for having built the pontoon bridge used by the king to cross the 'Thracian Bosporus' (c. 513–512 BCE): it displayed the bridge itself, and it was at once consecrated, according to Herodotus, by Mandrokles to the Heraion of Samos.[11] This gift was very political, emphasizing both Mandrokles' science and Darius I's power. When all is said and done, the political and honorary programme of Alexander's hearse was not so far from the one that was displayed by Darius I's painting.[12]

The great paintings of the fifth and fourth centuries BCE mention mostly 'historical' or 'mythical' representations.[13] So Polygnotus of Thasos (470–440 BCE), in the Lesche of the Knidians in Delphi, displayed the Iliupersis and the departure of the Greek fleet with a great deal of verisimilitude: 'On the ship of Menelaus they are preparing to put to sea. The ship is painted with children among the grown-up sailors; amidships is Phrontis the steersman holding two boat-hooks [...] beneath him is one Ithaemenes carrying clothes, and Echoeax is

---

10 Pausanias, 5.19.1. See Snodgrass 2001, 127–41.
11 Herodotus, 4.88. The same Herodotus mentions, during the siege of Phocaea by Harpagus, 'paintings' in the city (Herodotus, 1.164), without any precision. On Mandrokles: West 2013, 117–28. Many wooden votive offerings representing ships were found in that very sanctuary: Kyrieleis 1980, 89–94; Kyrieleis 1993, 99–122, sp. 112. These numerous offerings of the Archaic period were certainly related the marine cult of Hera: see Fenet and Jost 2016.
12 During the imperial period, Trajan also commissioned a representation of the bridge on the Danube. This work was conducted by his architect Apollodorus of Damascus, on his column including the bas-relief evidencing the conquest of Dacia (Coarelli 1999, 162, sc. 98–99). He also ordered carvings of several scenes of navigation, one of which showed him operating the 'rudder' of a warship (*idem*, 78 sc.34), while others represented two pontoon bridges on the Danube (sc. 3 and 47).
13 On this matter, see Hölscher 1973; Rouveret 1989, 129–61; and Linant de Bellefonds and Prioux 2017.

going down the gangway, carrying a bronze urn'.[14] His contemporary, Mikon, adorned the Stoa Poikile of Athens with a painting of the battle of Marathon, which mixed historical (Miltiades is emphasized)[15] with heroic characters (the heroes of Marathon, Theseus, Athena, Herakles and the hero Echetlaeus are displayed on the side of the Athenian *strategos*),[16] with 'the Phoenician ships, and the Greeks killing the foreigners who are scrambling into them'.[17]

In both cases, these paintings, displayed in symbolic locations, combined the representation of heroes and historical examples. Both used the reference to reality (technical or historical) to lend credence to an event that held great importance for the client for whom the work was made. And in both cases, the narration aimed to exalt civic values, namely those of the cities of Knidos and Athens. The most important aspect was not the documentary realism of the painting, but its visual verisimilitude, which heightened the fame of the artist because it enabled him to convince the spectator of the 'reality' of the painting and of the ideological programme it promoted.[18]

Another useful genre, which was employed early in the fifth century, was that of allegorical painting: the hero Marathon, displayed in Mikon's painting, is a good example of it. In the same period, Pausanias described a work by Panainos, Phidias' nephew (c. 450–430 BCE), which adorned the balustrade of the statue of Zeus in Olympia: there various heroes could be found, as well as '[…] Hellas, and Salamis carrying in her hand the ornament made for the top of a ship's bows'.[19] Portraits appear in parallel: Miltiades by Mikon, and work by Parrhasios (c.

---

14 Pausanias, 10.25.2. It is possible that Pausanias, who reads names inscribed on the table, has mistakenly identified the name *Echoiax* with one of the characters (Reinach 1921, reed. 1985, 93, note 3). About Polygnotus, see Cousin 1999, 61–103; and Hölscher 2015, 47–48.
15 Cornelius Nepos, *Miltiades* 6.3.
16 Pausanias, 1.16.
17 Pausanias, 1.15.
18 Hölscher 2015, 25: 'L'une des tâches fondamentales dévolues aux images consiste à rendre 'présents' des personnes, des objets ou des événements qui se trouvent être absents *in corpore*'. And Hölscher 2015, 51: 'De fait, toutes les déclarations des auteurs antiques portant sur l'art figuratif soulignent le caractère central de cette référentialité des images par rapport à la réalité'. Lastly Hölscher 2015, 53: 'S'il est vrai que l'image est une construction, *la réalité représentée dans l'image est également une construction*'.
19 Pausanias, 5.11.5.

420–370 BCE): 'He also painted [...] a Naval Commander in a Cuirass'.[20] Following the battle of Salamis in 480, the naval victory became a more and more popular theme.[21]

With the advent of Philip II and Alexander the Great, the question of the superiority of either 'history' or 'myth' was asked with more acuity. The conflict between the painter Apelles and the sculptor Lysippos is well known, the latter reproaching the former for having depicted, in one of his paintings, the hand of the king holding Zeus' lightning, whereas he (Lysippos) portrayed the Conqueror with a spear 'the glory of which no length of years could ever dim, since it was truthful and was his by right'.[22] One of their contemporaries, Nicias of Athens (350–300 BCE), likewise emphasized historical representation, leaving mythical subjects to the realm of poetry. And he may have mentioned, among the noble subjects of history, that of naval battles: 'The painter Nicias used to maintain that no small part of the painter's skill was the choice at the outset to paint an imposing object, and instead of frittering away his skill on minor subjects, such as little birds or flowers, he should paint naval battles and cavalry charges where he could represent horses in many different poses [...]. He held that the theme itself was a part of the painter's skill, just as plot was part of the poet's'.[23]

In the second century CE, Philostratus still praised the imitation of reality and explained that it was peculiar to painting: 'For imitation [...] in order to reproduce dogs, horses, humans, ships, everything under the sun'.[24] Actually, these kinds of painted subjects hardly evolved from Alexander's death to the time of imperial Rome, although after Actium more frequent representations of trade ships can be seen. This trend gained strength after Portus was founded by Claudius and the job of 'feeding' the plebs fell to the emperor. Few marine paintings have survived the great shipwreck of ancient works. At the very most two works by Pliny the Elder are worthy of mention. Theoros (c. 320–280 BCE) would have painted 'the Trojan War in a series of pictures';[25] and

---

20 Pliny the Elder, *Natural History* 35.69.
21 See Glasson 2014.
22 Plutarch, *Moralia, Isis and Osiris* 24.
23 Demetrius, *On Style* 76.
24 Philostratus, *Life of Apollonius* 2.22.
25 Pliny, *Natural History* 35.144 (representation that inspired the *Tabulae Iliacae* found in Rome?).

Nealkes of Sicyon (between the third and first century BCE), who: '[…] was a talented and clever artist, inasmuch as when he painted a picture of a naval battle between the Persians and the Egyptians, which he desired to be understood as taking place on the river Nile, the water of which resembles the sea, he suggested by inference what could not be shown by art: he painted an ass standing on the shore drinking, and a crocodile lying in wait for it'.[26] There is no doubt that there must have existed many others:[27] the Images by Philostratus, in the second century CE, provide an excellent example.[28]

4.2. Ship Painters

The activity of the painters who 'were adorning the ships' is more original. Since the Geometric period, ships were represented on Greek ceramics as having ornamental decoration on their bows:[29] a circle with crossed lines, which quickly evolved to become the well-known

---

26 Pliny, *Natural History* 35.142. It is worth noting that Aristotle, in *Problems* 23.6, laid down a pictorial rule that seems to have been followed: '[…] at any rate, painters paint rivers as pale, and the sea as blue.' In ancient art, representations of the Nile were mostly unaffected by this classification: see for example in the mosaic of Palestrina, where the waters of the river are shown in blue.
27 For example, Kydias of Kythnos (4th century BCE): '[…] for whose picture of the Argonauts the orator Hortensius paid 144,000 sesterces, and made a shrine for its reception at his villa at Tusculum.' (Pliny the Elder, *Natural History* 35.130.) The painting might have inspired the name of the portico of the Argonauts in Rome, built by Agrippa (Cassius Dio, 53.27: 'Meanwhile Agrippa beautified the city at his own expense. First, in honour of the naval victories he completes the building called the Basilica of Neptune and lent it added brilliance by the painting representing the Argonauts.' About Jason, Martial, *Epigrams* 11.1.12 speaks of the 'captain of the first ship', '*primae dominus carinae*'). See also Hippys (Hellenistic period?): 'Hippys for his Poseidon and his Victory.' (Pliny the Elder, *Natural History* 35. 141).
28 Philostratus, *Imagines* 1.19: 'Les Tyrrhéniens' (who took his inspiration from a myth of the *Homeric Hymn* to Dionysus): 'Now the pirate ship sails with warlike mien […] And, in order that it may strike terror into those they meet and they may look to them like some sort of monster, it is painted with bright colours, and it seems to see with grim eyes set into its prow, and the stern curves up in a thin crescent like the end of a fish's tail. As for the ship of Dionysus, it has a weird appearance in other respects, and it looks as if it were covered with scales at the stern, […] and its prow is drawn out in the semblance of a golden leopardess. Dionysus is devoted to this animal because it is the most excitable of animals and leaps lightly like a Bacchante.'
29 For example, a fragment of a Geometric krater of the Dipylon Master (A. Louvre, 517): https://www.louvre.fr/oeuvre-notices/cratere-fragmentaire; see in Basch 1987, 172, Fig. 353.

apotropaic eye.[30] These representations are frequent on Geometric and Orientalizing ceramics, and in the Attic works with black and red figures. This was the case with the naval battle that was represented on the Aristonothos krater (c. 675–650 BCE): one of the two ships bears an eye.[31] In the classical period, the ram often took an animal shape, for example a ram shaped like a boar's head,[32] or a fish's head.[33] The difficulty lies in understanding whether, in the image, this animal shape is constructed by the metallic part (the ram) or by various techniques with additions of figurative details painted on wooden and metal structural elements. In other words: whether the shape is intrinsic to the object or whether it is created with the use of paint.

Euripides, in *Iphigenia in Aulis*, suggests an interesting change to the 'Catalogue of Ships' by Homer. Although it takes place in a heroic time, the action of the play should have reminded the contemporaries of the tragedian of a familiar reality. However, for Euripides, the ornamentation of the ships is statuesque, spectacular, and rather located astern:

> 'I came to reckon and to behold
> their wondrous ships,
> to fill with pleasure
> the greedy vision of my female eyes.
> Holding the right flank
> of the fleet
> was the Myrmidon force from Phthia
> with fifty swift ships.
> In gilded images high upon their sterns
> stood Nereids,
> the ensign of Achilles' fleet.
> (…)

---

30 About the prophylactic eyes in marble found on the Agora in the Tektas Burnu shipwreck, in the harbour of Zea, see Carlson 2009, 347–65. The traces of paint that remain on a few of them reveal that they were certainly painted.

31 This vase was produced by a potter settled in Etruria: http://www.museicapitolini.org/fr/percorsi/percorsi_per_sale/museo_del_palazzo_dei_conservatori/sale_castellani/cratere_con_l_accecamento_di_polifemo_e_battaglia_navale).

32 About an Attic dinos (Basch 1987, 212, Fig. 440; or 217, Fig. 453; 227, Fig. 472: cup signed by Nikosthenes, Louvre, F 123: https://www.louvre.fr/oeuvre-notices/coupe-attique-de-type-figures-noires).

33 Basch 1987, 221, Fig. 460. See also the very beautiful ram-shaped vases from Apulia, for example the one which is conserved in the Petit Palais (Paris), inv. ADUT 422 (on this matter, Ambrosini 2010, 73–115: the creation of these large vases in Magna Graecia would have resulted from the introduction of the cult of Cybele in Southern Italy by Athens).

Next to them
with sixty ships from Athens,
was encamped
Theseus' son, who had the goddess Pallas
mounted on a chariot with winged steeds,
as the clear marker for his sailors.

The Boeotians' seagoing panoply,
fifty ships, I saw
blazoned with ensigns.
There was Cadmus
holding a golden serpent
aloft on the ships' high sterns.
Leitus, one of the Sown Men,
led his naval armament.
[…]
From Pylos I saw
Of Gerenian Nestor
[…]
the ensign upon his sterns, bull-footed in appearance,
the Alpheus River, his neighbor'.[34]

In the Hellenistic period, the painted ornamentation of the warships is documented by various representations such as the mosaics showing Berenike II's ram-shaped crown by Sophilos,[35] or the fresco found in Nymphaeum and displaying a 1.2-metre-long ship bearing the name Isis.[36] As for the Athlit ram, although it is not painted, it bears traces of its decoration.[37]

Ornaments in relief, rather than statuary, also appear on Greek coins, particularly those issued by the Lycian city of Phaselis, which, by the late fourth century, show staters with an eye at the bow and, above, the outline of a dolphin.[38] Another coin, from the same city, displays the same ram, but with the head of a gorgon, and the outline of

---

34 Euripides, *Iphigenia at Aulis* 231–76, trans. D. Kovacs.
35 Daszewski 1985, 142–58, pl. A, 32.
36 Basch 1985, 129–51.
37 Casson and Steffy 1991.
38 Item offered for sale in an auction in September 2016 (https://www.numisbids.com/n.php?p=lot&sid=2739&lot=18).

a cicada 'in front' of the ship.³⁹ Yet another coin from Phaselis represents a radiate bust (Helios) is laid on the deck, above the ram with an eye and a dolphin below;⁴⁰ or even a very beautiful detail of a ram with boar's head.⁴¹ After the conquest of the harbour by Ptolemy I in 309 BCE, busts of the king and of Berenice (with diadems) are represented above the ram, laid on the deck.⁴² In the same period, the tetradrachm of Demetrios Poliorcetes, commemorating the victory of Salamis in 306, shows a winged Victory blowing a trumpet on the bow of a ship. The work was often compared to the Victory of Samothrace, but it does not show an actual ship scene: these elements are above all political symbols, referring to the authority that issues the coins,⁴³ not necessarily depictions of real scenes. It is notable that these decorations are located at the bow, whereas Euripides located the ornaments astern: nevertheless, further documentary evidence indicates that the bow could indeed carry significant ornamentation.

One such example is the text written by Athenaeus of Naucratis, with its famous description of the Tessarakonteres, the giant ship of Ptolemy II Philopator, which specifies: 'Wonderful also was the adornment of the vessel besides; for it had figures at stern and bow not less than eighteen feet high, and every available space was elaborately covered with encaustic painting; the entire surface where the oars projected, down to the keel, had a pattern of ivy-leaves and Bacchic wands'.⁴⁴ The other ship belonging to the king, the Thalamegos, intended to sail along the Nile, included astern '[…] a frieze with striking figures in ivory, more than a foot and a half tall, mediocre in workmanship, to be sure, but remarkable in their lavish display. Over the dining-saloon was a beautiful coffered ceiling of cypress wood; the ornamentations on it were sculptured, with a surface of gilt'.⁴⁵

---

39 Item offered for sale in an auction in September 2016 (https://www.numisbids.com/n.php?p=lot&sid=2739&lot=18).
40 Bibliothèque nationale de France (http://catalogue.bnf.fr/ark:/12148/cb41780665q).
41 Bibliothèque nationale de France (http://catalogue.bnf.fr/ark:/12148/cb41780671n). This item would be dated to the fifth century BCE; see Basch 1987, 297, Fig. 626. On the coinage issued by Phaselis: Heipp-Tamer 1993.
42 Bibliothèque nationale de France (http://catalogue.bnf.fr/ark:/12148/cb41780658f).
43 On these representations, see most recently Badoud 2018, 279–306.
44 Athenaeus, *The Deipnosophists* 5.204 a-b.
45 Athenaeus, *The Deipnosophists* 5.205c.

Lastly, when discussing the giant ship of Hiero, Athenaeus mentions the external decoration: 'outside, a row of colossi, nine feet high, ran around the ship; these supported the upper weight and the triglyph, all standing at proper intervals apart. And the whole ship was adorned with suitable paintings.'[46] The 'figures' surrounded the ship on all sides: indeed, the size of the ship was similar to a floating palace.

Let us dwell for a moment on these 'painted paintings', these 'drawings made of wax' mentioned by Athenaeus. Ancient shipwrecks have not revealed much evidence of painted decoration.[47] Except the Marsala shipwrecks,[48] it is true that no shipwreck of a warship has been found. However, as Pliny the Elder explains, this kind of decoration was first lavished upon this kind of ship, surely because of its cost, and also because of the political significance of said decoration, which was not useful for a merchantman. Nevertheless, the prefect of the fleet at Misenum describes a change that occurred during his time: 'Wax is stained with these same colours for encaustic paintings, a sort of process which cannot be applied to walls but is common for ships of the navy (*classibus familiaris*), and indeed nowadays also for cargo vessels (*onerariis navibus*)[…]'.[49]

From a technical point of view, these paintings on a ship are therefore associated with encaustic painting. Although he does not provide any dates for this innovation, Pliny points out that this technique was first

---

46 Athenaeus, *The Deipnosophists* 5.208b. There was an inscription too: 'freshly charactered on its stout prow' (Athenaeus, *The Deipnosophists* 5.209d.). Rouveret (1989, 210–12) briefly mentions the ornamentation of these two giant ships. About these ships: Pomey and Tchernia 2006, 81–99; Castagnino Berlinghieri 2010, 169–88; Nantet 2016, 126–31.

47 Although very rare, traces of painting have been identified on very few ancient shipwrecks, such as the Herculanum shipwreck, whose hull revealed a white line. Steffy 1985, 519–21. The white line could have been a tonnage mark, even though this suggestion must be considered with caution, see Nantet 2016, 75, 430–31, E58. The Pisa shipwreck also showed some traces of painting: Colombini et al. 2003, 659–74. Dyes have been found in some shipwrecks, such as the La Madrague de Giens shipwreck (Liou and Pomey 1985, 564) and a few others (see references in the same article). They were used for the refection of the hull paintings. On the contrary, the dyes found on the Planier 3 shipwreck should be considered as part of the cargo (Tchernia 1968–1970, 51–82.). The Gyptis, a replica of the Jules-Verne 9 shipwreck, was painted: Pomey 2014, 1333–57. Pomey and Poveda 2018, 45–56. For photos, see Pomey and Poveda 2015.

48 Honor 1981.

49 Pliny the Elder, *Natural History* 35.49.

used on wax and ivory, with a *cestros*; 'later the practice of decorating battleships (*classes pingi*) was developed. There followed a third method, that of employing a brush when wax has been melted by fire; this process of painting ships (*quae pictura navibus*) is not spoilt by the action of the sun nor by saltwater or winds'.[50] Then Pliny specifies: 'A third of the white pigment is ceruse or lead acetate, the nature of which we have stated in speaking of the ores of lead. There was also once a native ceruse earth found on the estate of Theodotus at Smyrna, which was employed in old days for painting ships (*navium picturas*)'.[51]

The technique described was adapted to the environment for which the paintings were made, as were the materials used. This specialism, which was looked down upon, occurs on only two occasions in the history of Greek painting as established by Pliny the Elder: when two of those craftsmen managed to rise from the stigmatised occupation of ship painter to the rank of easel painter, and thus to make a name for themselves as artists.

The first was Protogenes of Caunus (300–240 BCE), from Rhodes. Apelles' contemporary, he experienced rough start:

> At the outset he was extremely poor, and extremely devoted to his art and consequently not very productive. The identity of his teacher is believed to be unrecorded. Some people say that until the age of fifty he was also a ship painter (*naves pinxisse*), and that this is proven because when, later in life, he was decorating the gateway of the Temple of Athene on a very famous site in Athens, (where he created his famous *Paralus*

---

50   Pliny the Elder, *Natural History* 35.149. See also Pliny the Elder, *Natural History* 16.56: 'We must not omit to state that among the Greeks also the name of 'live pitch' [*zopissa*] is given to pitch that has been scraped off the bottom of seagoing ships and mixed with wax — as life leaves nothing untried — and which is much more efficacious for all the purposes for which the pitches and resins are serviceable, this being because of the additional hardness of the sea salt.' Pliny the Elder, *Natural History* 24.26: '*Zopissa*, as I have said, is scraped off ships, wax being soaked in sea brine. The best is taken from ships after their maiden voyage. It is also added to poultices to disperse gatherings.' These lines reveal a few details about the technical operations of ship maintenance. On the use of pitch: Connan et al. 2002, 177–96; and Connan 2002, 2–9, who mention a mixing of pitch and wax that covered the surface of the planking of the Archaic hull of the Jules-Verne 9 shipwreck. More recently, pitch has been identified on the Arles-Rhône 3 shipwreck, as it was used for luting, cf. Marlier 2014, 115–16.

51   Pliny the Elder, *Natural History* 35.19. There would have been a confusion here with the *cresa viridis* (Dauzat 1997, 35, note 77). On wax as a technique used in painting and carving, see Bourgeois 2014, 69–80.

and *Hammonias*, which is also called the *Nausicaa* [two sacred ships of Athens] by some people), he added some small drawings of battleships in what painters call the 'side-pieces,' (*parergia*) in order to show from what origins his work had come, to arrive at the pinnacle of this glorious display.[52]

The location of this painting in the Propylaea, and the subject (the sacred *triereis* of Athens), show how significant Protogenes had become by the first half of the third century. One interpretation of Pliny's words is that Protogenes was painting, not ships, but *ex-votos* representing ships. It is true that this practice existed in Antiquity,[53] as evidenced by Latin sources. What was it?

Cicero clearly mentions the votive tablets that were offered in order to express gratitude to the protective deity after a storm:

> You object that on occasion good men achieve successes; indeed, we latch on to those, and without any justification attribute them to the immortal gods. The opposite was the case when Diagoras, whom they call the Atheist, visited Samothrace, where a friend remarked to him: 'You believe that the gods are indifferent to human affairs, but all these tablets (*tabulis pictis*) with their portraits surely reveal to you the great number of those whose vows enabled them to escape the violence of a storm, so that they reached harbor safe and sound.' 'That is the case', rejoined Diagoras, 'but there are no portraits (*picti*) in evidence of those who were shipwrecked and drowned at sea'.[54]

This practice is confirmed by Juvenal, who insists on its importance[55] and describes another practice, associated this time with begging: 'The person [...] will now have to be satisfied with rags covering his freezing crotch and with scraps of food while he begs for pennies as a shipwreck survivor and maintains himself by painting a picture (*picta*) of the storm'.[56]

Was Protogenes a painter who specifically produced marine paintings?[57] The expression used by Pliny (*naves pingere*) is not the same

---

52 Pliny the Elder, *Natural History* 35.101. About the Athenian warships, see Bubelis 2010, 385–411, *Historia* 59.4.
53 And much later: Rieth and Milon 1981.
54 Cicero, 3.89, trans. Walsh 1997.
55 Juvenal, 12.25: '[...] a different kind of danger. Listen and pity him a second time. The rest is, admittedly, part of the same experience, terrible without doubt, but familiar to many, as all those shrines with their votive tablets (*fana tabella plurima*) indicate. Everyone knows that painters make their bread and butter from Isis.'
56 Juvenal, 14.300.
57 This was Reinach's interpretation (1985, 399) of Pliny's two texts dealing with the 'ship painters'. Likewise, de Ridder (1915, 282–87), who connected in a series the

as Cicero's or Juvenal's (who mention *tabula* or *tabella*):⁵⁸ the support of Protogenes' paintings was therefore the ship herself, similar to the paintings of Herakleides the Macedonian (who lived around 168 BCE): 'Heraclides of Macedon is also a painter of note. He began by painting ships (*initio naves pinxit*), and after the capture of King Perseus he migrated to Athens…'.⁵⁹

As we know, the practice of painting with wax did not disappear.⁶⁰ Indeed, the arrival of the ship of Cybele, Mother Goddess, in Rome, in 204 BCE, was celebrated two centuries later by Ovid: '[…] a thousand hands assemble, and the Mother of the Gods is lodged in a hollow ship painted in encaustic colours (*picta coloribus ustis*)'.⁶¹

4.3. Conclusion

The end of the Hellenistic period merged with the Roman period. A few artists perpetuated the maritime and historical paintings, like Androbios,⁶² while mural paintings developed too:⁶³

> […] Spurius Tadius also, during the period of his late lamented Majesty Augustus, was cheated of his due, who first introduced the most attractive fashion of painting walls with pictures of country houses and porticoes […] rivers, coasts, and whatever anybody could desire, together with various sketches of people going for a stroll or sailing in a boat […] people fishing […]'.⁶⁴

As in Greek verse, Latin poetry commemorates some memorable ship battles, such as Propertius who describes the battle of Actium:

> Nor let it frighten you that their armada sweeps the waters with many hundred oars: the sea o'er which it glides likes it not. And all the Centaurs

---

      ram-shaped bas-relief found in Rhodes (acropolis of Lindos) and the funerary steles of Rhenea.
58   Plisecka 2011.
59   Pliny the Elder, *Natural History* 35.135.
60   See La Torre et al. 2011; Linant de Bellefonds et al. 2015.
61   Ovide, *Fasti* 5.275–76.
62   Pliny the Elder, *Natural History* 35.138: 'Androbius painted a Scyllus Cutting the Anchor-ropes of the Persian Fleet.'
63   Examples include the mural paintings of the temple of Isis in Pompei, currently displayed in the Archaeological Museum of Naples.
64   Pliny the Elder, *Natural History* 35.116–117. On Roman painting in general (including Pliny), see Croisille 2005; and on the Roman collectors: Routledge 2012.

threatening to throw rocks borne by their prows will prove to be naught but hollow planks and painted scares.[65]

This ship decoration can be found on the marble frieze from the time of Claudius, evoking the battle of Actium, which shows Antony's ship with a rearing Centaur as a figurehead (conversely, the *Scylla* of Augustus' ship would have disappeared during a modern restoration).[66] The Vatican Virgil still shows Aeneas's ships with statues at their bow.[67] As for the bronze ornaments of the Roman ships of Nemi, they are to a certain extent the monumental heirs of the wax figures of the Hellenistic 'ship painters'.[68]

Bibliography

Primary Sources

Manuscripts

Vaticanus Lat. 3225 (Vergil, Opera).' https://digi.vatlib.it/view/MSS_Vat.lat.3225/0001

Literary sources

Babbitt, F. C., trans. 1969. *Plutarch. Moralia*. Vol. 5. *Isis and Osiris*. Cambridge, MA, London: Loeb Classical Library. Original edition, Cambridge, MA, London: Loeb Classical Library, 1936.

Braund, S. M., ed. and trans. 2004. *Juvenal. The Satires*. Cambridge, MA, London: Loeb Classical Library.

Cary, E., ed. and trans. 1955. *Cassius Dio. Roman history*. Vol. 6. Cambridge, MA, London: Loeb Classical Library. Original edition, Cambridge, MA, London: Loeb Classical Library, 1917.

---

65 Propertius, *Elegies* 4.6.50 and the following verses.
66 Copy of the Medinaceli-Budapest bas-reliefs (Réunion des musées nationaux et du Grand Palais des Champs-Élysées 2014, 292–95); see also Tomei 2017 (http://journals.openedition.org/mefra/4446): marble fragments show ships decorated with various carved characters. A gem conserved in the Berlin museum reveals a huge ship with the figure of a bull on her bow: was it a reminder of the Hiero's Syracusia? (Basch 1987, 471, 474. Fig. 1070).
67 Vatican, Latin manuscript 3225, folio XLII recto and XLIII verso (https://digi.vatlib.it/view/MSS_Vat.lat.3225/0001).
68 Lastly on these ornaments: Wolfmayr 2013; Frielinghaus et al. 2017, 91–104.

Rolfe, J. C., trans. 1929. *Cornelius Nepos. The Book on the Great Generals of Foreign Nations*. Cambridge, MA, London: Loeb Classical Library.

Evelyn-White, H. G., trans. 1959. *Hesiodos. The Shield of Heracles*. Cambridge, MA, London: Loeb Classical Library. Original edition, Cambridge, MA, London: Loeb Classical Library, 1914.

Fairbanks, A., trans. 1960. *Philostratus. Imagines*. Cambridge, MA, London: Loeb Classical Library. Original edition, Cambridge, MA, London: Loeb Classical Library, 1931.

Geer, R. M., ed. and trans. 1947. *Diodorus Siculus*. Vol. 9. *The Library of History*. Cambridge, MA, London: Loeb Classical Library.

Douzat, P., ed. 1997. *Pline l'Ancien, Histoire Naturelle*. Paris: Les Belles Lettres.

Godley, A. D., trans. 1928. *Herodotus*. Vol. 1. Cambridge, MA, London: Loeb Classical Library. Original edition, Cambridge, MA, London: Loeb Classical Library, 1921.

Goold, G. P., ed. and trans. 1990. *Propertius. Elegies*. Cambridge, MA, London: Loeb Classical Library.

Gulick, C. B., ed. and trans. 1957. *Athenaeus. The Deipnosophists*. Vol. 2. Cambridge, MA, London: Loeb Classical Library. Original edition, Cambridge, MA, London: Loeb Classical Library, 1928.

Innes, D. C. ed. and trans. 1995. *Demetrius. On Style*. Cambridge, MA, London: Loeb Classical Library (based on W. Rhys Roberts).

Jones, C. P., ed. and trans. 2005. *Philostratus*. Vol. 1. *The Life of Apollonius of Tyana*. Cambridge, MA, London: Loeb Classical Library.

Jones, W. H. S., trans. 1959. *Pausanias. Description of Greece*. Vol. 1. Cambridge, MA, London: Loeb Classical Library. Original edition, Cambridge, MA, London: Loeb Classical Library, 1918.

Jones, W. H. S., trans. 1965. *Pausanias. Description of Greece*. Vol. 4. Cambridge, MA, London: Loeb Classical Library. Original edition, Cambridge, MA, London: Loeb Classical Library, 1935.

Jones, W. H. S., trans. 1966. *Pliny the Elder. Natural History*. Vol. 7. Cambridge, MA, London: Loeb Classical Library. Original edition, Cambridge, MA, London: Loeb Classical Library, 1956.

Jones, W. H. S., and h. A. Ormerod, trans. 1966. *Pausanias. Description of Greece*. Vol. 2. Cambridge, MA, London: Loeb Classical Library. Original edition, Cambridge, MA, London: Loeb Classical Library, 1926.

Kovacs, D., ed. and trans. 2002. *Euripides*. Vol. 6. *Iphigenia at Aulis*. Cambridge, MA, London: Loeb Classical Library.

Mayhew, R., ed. and trans. 2011. *Aristotle*. Vol. 9. *Problems*. Cambridge, MA, London: Loeb Classical Library.

Murray, A. T., trans. 1965. *Homer. The Iliad.* Vol. 1. Cambridge, MA, London: Loeb Classical Library. Original edition, Cambridge, MA, London: Loeb Classical Library, 1924.

Murray, A. T., trans. 1966. *Homer. Odyssey.* Vol. 1. Cambridge, MA, London: Loeb Classical Library. Original edition, Cambridge, MA, London: Loeb Classical Library, 1919.

Rackham, H., trans. 1952. *Pliny the Elder. Natural History.* Vol. 9. Cambridge, MA, London: Loeb Classical Library.

Rackham, H., trans. 1968. *Pliny the Elder. Natural History.* Vol. 4. Cambridge, MA, London: Loeb Classical Library. Original edition, Cambridge, MA, London: Loeb Classical Library, 1945.

Shackleton Bailey, D. R., ed. and trans. 1993. *Martial. Epigrams.* Vol. 2. Cambridge, MA, London: Loeb Classical Library.

Frazer, J. G., trans. 1967. *Ovid. Fasti.* Cambridge, MA, London: Loeb Classical Library. Original edition, Cambridge, MA, London: Loeb Classical Library, 1931.

Walsh, P. G., trans. 1997. *Cicero. The Nature of the Gods.* Oxford: Clarendon Press.

Waterfield, R., trans. 1998. *Herodotus. The Histories.* Oxford, NY: Oxford University Press.

Ceramic sources

'Coupe attique de type A à figures noires.' (Kardianou-Michel, A.) *Louvre*. https://www.louvre.fr/oeuvre-notices/coupe-attique-de-type-figures-noires

'Cratère fragmentaire.' (Kardianou-Michel, A.) *Louvre*. https://www.louvre.fr/oeuvre-notices/cratere-fragmentaire

'Cratere con l'accecamento di Polifemo e battaglia navale.' *Musei Capitolini*. http://www.museicapitolini.org/it/percorsi/percorsi_per_sale/museo_del_palazzo_dei_conservatori/sale_castellani/cratere_con_l_accecamento_di_polifemo_e_battaglia_navale

'Monnaie: Statère, Argent, Phasélis, Lycie. No. FRBNF41780665. IFN-8524144.' *BNF*. http://catalogue.bnf.fr/ark:/12148/cb41780665q

'Monnaie: Statère, Argent, Phasélis, Lycie. No. FRBNF41780671. IFN-8524150.' *BNF*. http://catalogue.bnf.fr/ark:/12148/cb41780671n

'Monnaie: Statère, Argent, Phasélis, Lycie. No. FRBNF41780658. IFN-8524137.' *BNF*. http://catalogue.bnf.fr/ark:/12148/cb41780658f

'Statère – Phaselis (4ème siècle av. J. C.).' *Numisbids*. https://www.numisbids.com/n.php?p=lot&sid=2739&lot=18

Tomei, M. A. 2017. 'Il monumento celebrativo della battaglia di Azio sul Palatino.' *MEFRA* 129 (2). http://journals.openedition.org/mefra/4446

Secondary Sources

Ambrosini, L. 2010. 'Sui vasi plastici configurati a prua di nave (trireme) in ceramica argentata e a figure rosse.' *MEFRA* 122 (1):73–115. https://doi.org/10.4000/mefra.336

Badoud, N. 2018. 'La Victoire de Samothrace, défaite de Philippe V.' *RA* 66 (2):279–306. https://doi.org/10.3917/arch.182.0279

Basch, L. 1985. 'The Isis of Ptolemy II. Philadelphus. ' *Mariner's Mirror* 2(71): 129–51. https://doi.org/10.1080/00253359.1985.10656020

Basch, L. 1987. *Le Musée imaginaire de la marine antique.* Athènes: Institut Hellénique pour la Préservation de la Tradition Nautique.

Berlinghieri, C. E. F. 2010. 'Archimede alla corte di Hierone II: dall'idea al progetto della della più grande nave del mondo antico, la Syrakosia.' *Hesperia 26, Studi sulla grecità di Occidente*: 169–88.

Bourgeois, B. 2014. '(Re)peindre, dorer, cirer: la *thérapéia* en acte dans la sculpture grecque hellénistique.' *Technè* 40: 69–80.

Bubelis, W. 2010. 'The sacred Triremes and their *tamiai* at Athens.' *Historia* 59(4): 385–411.

Carlson, D. 2009. 'Seeing the Sea: Ships' Eyes in Classical Greece.' *Hesperia* 78: 347–65. https://doi.org/10.2972/hesp.78.3.347

Casson, L. 1964. 'Odysseus' boat (Od. V, 244–53).' *AJP* 85:61–64.

Casson, L. 1992. 'Odysseus' boat (Od. V, 244–53).', *IJNA* 21:73–74.

Casson, L., and J. R. Steffy, eds. 1991. '*The Athlit ram.*' College Station: Texas A&M University Press.

Coarelli, F. 1999. *La colonna Traiana.* Rome: Istituto Archeologico Germanico Anno.

Colombini, M. P., G. Giachi, F. Modugno, P. Pallecchi, and E. Ribechini. 2003. 'Characterisation of paints and waterproofing materials of the shipwrecks found in the archaeological site of the Etruscan and Roman Harbour of Pisa (Italy).' *Archaeometry* 45(4): 659–74. https://doi.org/10.1046/j.1475-4754.2003.00135.x

Connan, J. 2002. 'Le calfatage des bateaux.' *Pour la Science* 298: 2–9.

Connan, J., B. Maurin, L. Long and H. Sebire. 2002. 'Identification de poix et de résine de conifère dans des échantillons archéologiques du lac de Sanguinet: exportation de poix en Atlantique à l'époque gallo-romaine.' *Revue d'Archéométrie* 26: 177–96. https://doi.org/10.3406/arsci.2002.1032

Cousin, C. 1999. 'Composition, espace et paysage dans les peintures de Polygnote à la *leschè* de Delphes.' *Gaia* 4: 61–103.

Croisille, J.-M. 2005. *La Peinture romaine.* Paris: Picard.

De Ridder, A. 1915. 'Protogène' *REG* 128–129: 282–87.

Daszewski, W. 1985. 'Corpus of Mosaics from Egypt. Hellenistic and Early Roman Period.' *Aegyptiaca Treverensia* 3 (1): 142–58.

Fenet, A., and M. Jost. 2016. *Les Dieux olympiens et la mer: espaces et pratiques cultuelles. Collection de l'École française de Rome* 509. Rome: École française de Rome. https://doi.org/10.4000/books.efr.5580

Frost, H., ed. 1981. *Lilybaeum (Marsala) The Punic Ship: Final excavation report. Atti della Accademia nazionale dei Lincei. Notizie degli scavi di antichità* Suppl. of vol. 30 (1976). Rome: Accademia nazionale dei Lincei.

Glasson, P. 2014. 'Les représentations de la victoire navale de la haute époque hellénistique à Auguste.' PhD. diss., Université Paris-Sorbonne (Paris IV).

Heipp-Tamer, C. 1993. *Die Münzprägung der lykischen Stadt Phaselis in griechischer Zeit. Saarbrücker Studien zur Archäologie und Alten Geschichte* 6. Saarbrücken: Saarbrücker Druckerei.

Hölscher, T. 1973. *Griechische Historienbilder des 5. Und 4. Jahrhunderts v. Chr. Beiträge zur Archäologie* 6. Würzburg: Triltsch.

Hölscher, T. 2015. *La Vie des images grecques. Sociétés de statues, rôles des artistes et notions d'esthétiques dans l'art grec ancient*. Paris: Hazan.

Kyrieleis, H. 1980. 'Archaische Holzfunde aus Samos.' *MDAI(A)* 95: 89–94.

Kyrieleis, H. 1993. 'The Heraion at Samos.' In *Greek Sanctuaries. New Approaches*, edited by N. Marinatos and R. Hägg, 99–122. New York: Routledge.

La Torre, G. F., and M. Torelli. 2011. *Pittura ellenistica in Italia e in Sicilia: linguaggi e tradizioni*, Actes du colloque de Messine 24–25 septembre 2009, *Archaeologica*, 163, Rome: Giorgio Bretschneider Editore.

Linant de Bellefonds, P., E. Prioux and A. Rouveret, eds. 2015. *D'Alexandre à Auguste: dynamiques de la création dans les arts visuels et la poésie*. Rennes: Presses Universitaires de Rennes.

Linant de Bellefonds, P., and E. Prioux. 2017. *Voir les mythes: poésie hellénistique et arts figures*. Paris: Picard.

Liou, B., and P. Pomey. 1985. 'Recherches archéologiques sous-marines.' *Informations archéologiques, Gallia* 43 (2): 547–76.

Mark, S. E. 1991. 'Odyssey 5.234–53 and Homeric Ship Construction: A Reappraisal.' *AJA* 95: 441–45.

Mark, S. E. 1996. 'Odyssey 5. 234–53 and Homeric ship construction: a clarification.' *IJNA* 25:46–48.

Mark, S. E. 2005. *Homeric Seafaring*. College Station: Texas A&M University Press.

Marlier, S. 2014. 'L'épandage de poix.' In *Arles-Rhône 3. Un chaland gallo-romain du Ier siècle après Jésus-Christ, Archaeonautica* 18, edited by S. Marlier, 115–16. https://doi.org/10.3406/nauti.2014.1316

Réunion des musées nationaux et du Grand Palais des Champs-Élysées. 2014. *Auguste.* Paris: Réunion des musées nationaux — Grand Palais (exhibition catalogue 2013–2014).

Nantet, E. 2016. 'Phortia. Le tonnage des navires de commerce en Méditerranée du VIIIe siècle avant l'ère chrétienne au VIIe siècle de l'ère chrétienne.' Rennes: Presses Universitaires.

Plisecka, A. 2011. *Tabula picta: Aspetti giuridici del lavoro pittorico in Roma antica.* Padua: CEDAM.

Pomey P., 2014. 'Le projet Prôtis. Construction de la réplique navigante d'un bateau grec du VIe siècle av. J.-C.' *Comptes Rendus Académie des Inscriptions et Belles-Lettres* 3 (juillet–octobre): 1333–57.

Pomey P., and P. Poveda. 2018. 'Gyptis: Sailing Replica of a 6th-century-BC Archaic Greek Sewn Boat.' *International Journal of Nautical Archaeology* 47(1): 45–56.

Pomey P., and P. Poveda 2015. *Le Gyptis, Reconstruction d'un navire antique. Notes photographiques Marseille (1993–2015).* Paris: CNRS Éditions.

Pomey, P., and A. Tchernia. 2006. 'Les inventions entre l'anonymat et l'exploit: le pressoir à vis et la Syracusia.' In *Innovazione tecnica e progresso economico nel mondo romano, Atti degli incontri capresi di storia dell'economia antica (Capri: 2003),* edited by E. Lo Cascio, 81–99. Bari, Edipuglia.

Reinach, A. 1921. *Textes grecs et latins relatifs à la peinture ancienne: recueil Milliet.* Paris: Klincksieck. Reedited by A. Rouveret. 1985. Paris: Macula.

Rieth, É., and A. Milon. 1981. *Ex Voto marins dans le monde: de l'Antiquité à nos jours (catalogue d'exposition du Musée de la Marine).* Paris: Musée National de la Marine.

Rouveret, A. 1989. *Histoire et imaginaire de la peinture ancienne (Ve siècle av. J.-C. – Ier siècle ap. J.-C.). (BEFAR 274).* Rome: École française de Rome.

Rouveret, A. 2017. 'Adolphe Reinach (1887–1914): peinture antique et modernité.' In *Au-delà du savoir: les Reinach et le monde des arts. Cahiers de la villa Kérylos* 28, edited by J. Jouanna, H. Lavagne and A. Pasquier, 61–84. Paris: Éditions de Boccard.

Rutledge, S. H. 2012. *Ancient Rome as a Museum. Power, Indentity, and Culture of Collecting. Oxford Studies in Ancient Culture & Representation.* Oxford: Oxford University Press. https://doi.org/10.1093/acprof:oso bl/9780199573233.001.0001

Snodgrass, A. 2001. 'Pausanias and the Chest of Kypselos.' In *Pausanias: Travel and Memory in Roman Greece,* edited by S. E. Alcock, J. Cherry and

J. Elsner, 127–41. Oxford: Oxford University Press. https://doi.org/10.3366/edinburgh/9780748623334.003.0023

Steffy, J. R. 1985. 'The Herculaneum Boat: preliminary notes on hull details.' *AJA* 89: 519–21.

Tchernia, A. 1968–1970. 'Premiers résultats des fouilles de juin 1968 sur l'épave 3 de Planier.' *Etudes Classiques*, 3: 51–82.

Tchernia, A. 2001. 'Eustache et le rafiot d'Ulysse (Od. V).' In *Technai: techniques et sociétés en Méditerranée*, edited by J.-P. Brun and P. Jockey, 625–31. Paris: Maisonneuve & Larose.

West, S. 2013. 'Every Picture tells a Story.' In *Herodots Quellen. Die Quellen Herodots. Classica et Orientalia* 6, edited by B. Dunsch, K. Ruffing and K. Dross-Krüpe, 117–28. Wiesbaden: Harrassowitz Verlag.

Wolfmayr, S. 2017. 'Über die Bronzefunde der Nemisee-Schiffe.' In *Schiffe und ihr Kontext: Darstellungen, Modelle, Bestandteile: von der Bronzezeit bis zum Ende des Byzantinischen Reiches. Actes du colloque des 24–25 Mai 2013*, edited by H. Frielinghaus, Th. Schmidts and V. Tsamakda, 91–104. Mayence: Schnell & Steiner.

5. The Rise of the Tonnage in the Hellenistic Period

Emmanuel Nantet

During the Hellenistic period, various kinds of evidence demonstrate the existence of many large ships whose tonnage was greater than one hundred tonnes and could even reach several hundred tonnes. However, the accurate evolution of their tonnage is more complicated to determine for ships that sailed in the Eastern Mediterranean. It seems that an initial increase occurred in the first part of the second century BCE all over the Mediterranean. This particularly affected wheat and stone, since these goods required large ships. The increase in tonnage during this period was due to a desire for more prestige, influenced by political and military factors, and less to do with a desire for increased profits.

A second rise seems to have occurred from the end of the second century to the beginning of the first century BCE. However, this growth was restricted to particular routes in the Mediterranean, and only to ships carrying very valuable merchandise, such as wine or works of art. In fact, the development in tonnage was obviously the result of the significant changes in maritime trade caused by Roman rule.

Beyond these factors, the growth in tonnage during the Hellenistic period is due to developments in both ship and harbour technologies. Of course, the economic background — the growth of cities in the Hellenistic world — helped stimulate the demand for big ships.

From the end of the sixth to the fourth century BCE, tonnage increased considerably. In the Archaic period, tonnage was limited to a few dozen tonnes,[1] but it reached one hundred tonnes and even more in the Classical period. How did the situation develop in the Hellenistic period? Was it the same in the entire Mediterranean? Did changes in tonnage occur when Rome took control of the Eastern Mediterranean?

5.1. The Sources

Shipwrecks are our most accurate sources for answering these questions. However, the tonnage of most shipwrecks is often difficult to determine. Only the shipwrecks whose tonnage can be calculated according to the three methods suggested by Patrice Pomey can be compared.[2] Currently, we have results from eighteen shipwrecks matching Pomey's criteria (Table 5.1). Thus, the evolution of tonnage in this period can be represented on a graph (Fig. 5.1).

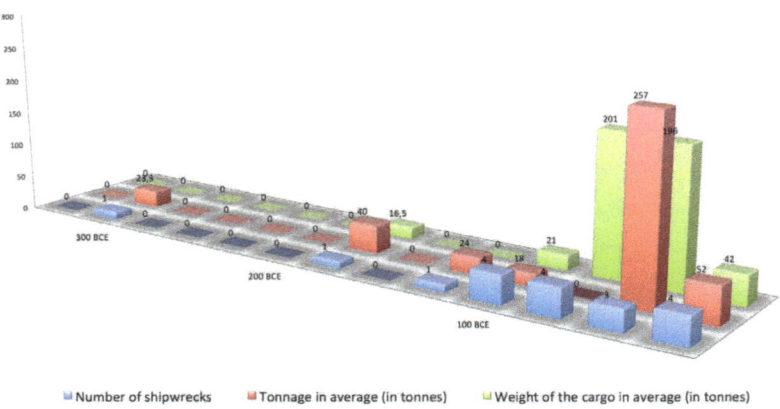

Figure 5.1 The evolution of the tonnage of the ships in the Hellenistic period from the shipwrecks. Graph by Emmanuel Nantet. CC BY.

---

1 Nantet 2016.
2 Pomey and Rieth 2004; Nantet 2016, 2017.

5. The Rise of the Tonnage in the Hellenistic Period

Table 5.1 The tonnage of the ships in the Hellenistic period from the shipwrecks (shipwrecks found in the Eastern Mediterranean are marked in grey).

Name of the shipwreck	Location	Dates (BCE)	Deadweight tonnage (in tonnes)	Weight of the cargo (in tonnes)	Method of estimation (Pomey and Rieth, 2004)	Cargo	No. (Nantet 2016)
Mazotos	Cyprus	350/325	-	20	Weight of the cargo	500 amphorae	19
Kyrenia	Cyprus	295/285	23,3	-	Hull lines reconstruction	404 amphorae, 10,000 almonds	20
Chrétienne C	France	175/150	40	16,5	Tonnage formula	500 amphorae	21
Apollonia 1	Libya	150/80	24	-	Tonnage formula	-	22
Carry-le-Rouet	France	125/75	-	25	Weight of the cargo	24 slabs	23
Dramont C	France	125/75	14	8,2	Tonnage formula	130 amphorae, app. 50 iron bars, resin, 3 millstones, ballast stones	24
La Ciotat 3	France	125/75	-	40	Weight of the cargo	One thousand amphorae, common ware	25
Cavalière	France	app. 100	22,17	12,43	Hull lines reconstruction	25 amphorae, pork, pottery, ballast stones	26
Albenga	Italy	100/80	-	500-600	Weight of the cargo	11,000 to 13,500 amphorae	27
Mahdia	Tunisia	100/80	-	230-250	Weight of the cargo	Slabs, works of art	28
Bénat 2	France	125/50	-	3,3	Weight of the cargo	3 *dolia*, amphorae, common ware	29
Le Miladou	France	125/50	-	11,25	Weight of the cargo	250 amphorae	30
La Madrague de Giens	France	75/60	402,5	320-350	Hull lines reconstruction	5,000 à 6,500 amphorae, pottery, sand ballast	31
Kızılburun	Turkey	100/25	-	54-61	Weight of the cargo	9 slabs, 24 amphorae, pottery	32
Dramont A	France	app. 50	111	-	Tonnage formula	Amphorae, ballast stones	33
Planier 3	France	50/47	32-46	-	Tonnage formula	Amphorae, dyes	34
Le Titan	France	50/30	60-70	58,7	Weight of the cargo	1,700 amphorae	35
Cap Béar 3	France	40/30	-	9,15	Weight of the cargo	260 amphorae	36

Of course, it is debatable how representative this list is of the historical reality. It does not include some well-known shipwrecks, such as the Antikythera shipwreck,[3] but the tonnage of the latter is extremely uncertain.

Most of the shipwrecks included in the list are located in the Western Mediterranean. We only know a few Hellenistic shipwrecks in the Eastern Mediterranean whose tonnage can be estimated,[4] such as Mazotos,[5] Kyrenia,[6] Apollonia 1 and Kızılburun.[7] Nevertheless, it

---

3   Weinberg et al. 1965; Christopoulou et al. 2012; Kaltsás et al. 2012.
4   The next four shipwrecks are briefly presented in Nantet 2016, n° 19, 20, 22 and 32.
5   Demesticha 2011.
6   Steffy 1985.
7   Carlson and Aylward 2010.

should be emphasised that some of the most interesting shipwrecks from the Late Republican period that lie in Western waters actually originated from ports in the Aegean Sea, such as the Mahdia shipwreck.[8] This archaeological evidence must be compared with evidence from papyri. Indeed, the Ptolemaic administration produced an abundance of documentation in order to manage the grain supply of Alexandria.[9] These papyri often mention the tonnages of the ships involved in this enormous task.[10] It seems that most of the ships mentioned in the papyri, such as the *kerkouroi*, studied by Pascal Arnaud, were sailing not only along the Nile, but also in the Mediterranean.[11] Papyri have been collected that document the period between the third and the first centuries BCE; they mention eighty-eight different tonnages.[12] The evolution of the tonnage can thus be represented on another graph (Fig. 5.2).

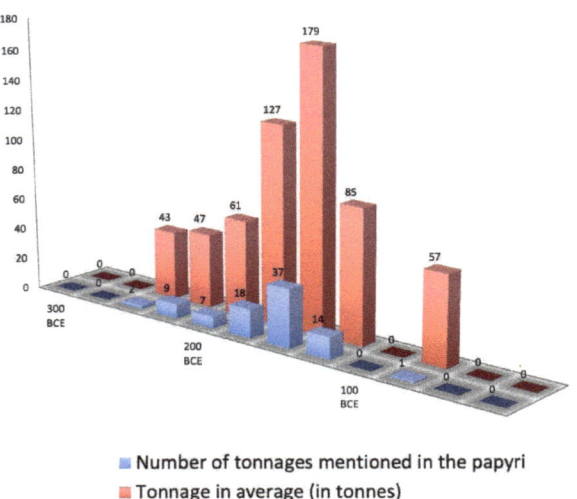

Figure 5.2 The evolution of the tonnage of the boats mentioned in the papyri in the Hellenistic period. Graph by Emmanuel Nantet. CC BY.

---

8   Hellenkemper Salies 1994.
9   Thompson 1983.
10   Hauben 1971, 1978, 1997; Meyer-Termeer 1978; Nantet 2016, 574–84.
11   Arnaud 2015.
12   For a full list of the papyri used for this study, see Nantet 2016, 575–79.

5. The Rise of the Tonnage in the Hellenistic Period 79

The epigraphical evidence, even though it relies only on a few documents, includes an inscription from Thasos about harbour regulations, first published by Marcel Launey.[13] The shape of the letters shows that this *stela* must have been produced during the third century BCE.

> Harbour regulation of Thasos
> (Third century BCE)
> IG XII, Suppl. 348
> Launey 1933, ed.
>
> 1 [πλ]οῖον μὴ [ἀ]νέλκειν ἐν τοῖς τῶν ..ργ... τοῦ μὲν πρώτου ἐλά[σσω φορ]-
> [τον ἄγον τρ]ισχι[λ]ίων ταλάντων, τοῦ <δὲ> δευτέρο[υ] ἐλάσσω ἄγο[ν] πεντα[κ]ισ-
> χ[ιλίων] τα[λάντω]ν.
>
> It is forbidden to haul a ship inside the limits, the first ones if the ship has a capacity of less than 3000 talents [about 61 tonnes], the second ones if the ship has a capacity of less than 5000 talents [about 102 tonnes].

This regulation reveals that there were three parts to the harbour of Thasos:[14]

- one for the ships of lower tonnage, of a capacity less than 3000 talents, or 61 tonnes;
- one for the ships of average tonnage, of a capacity between 3000 and 5000 talents, or 102 tonnes;
- one for the ships of larger tonnage, of a capacity beyond 5000 talents, or 102 tonnes.

The inscription was studied by several scholars.[15]

These sources provide evidence of the evolution in tonnage, which increased during this period. Nevertheless, it is hard to say whether

---

13 Launey 1933.
14 It is assumed here that a talent weighed 20.46 kg, see Nantet 2016, 573, note 29. Note that the level of 3000 talents has been debated because it is not legible. But it seems to be the most relevant suggestion.
15 Casson 1971; Houston 1988.

this was a continuous rise or if it can be broken down into several sequences — although it seems possible to distinguish at least two different episodes.

5.2. An Initial Rise in the First Part of the Second Century?

During the Early Hellenistic period, the tonnages mentioned in the papyri seem to be low enough. Only three *kerkouroi* that were mentioned ranged from 5000 *artabae* (114 tonnes) to 10,000 *artabae* (227 tonnes) — most of the tonnages remained at less than 5000 *artabae* during most of the third century BCE. This is more or less the same as the tonnage of the ships used in Athens to transport grain during the fourth century.[16] It corresponds to higher level named in the third-century harbour *stela* of Thasos (5000 talents, i.e. 102 tonnes), which allowed the bigger ships to moor in a deeper basin. Thus, in the third century, the tonnages would have been quite similar to those of the Classical period.

However, it appears that an increase occurred in the first half of the second century BCE, despite the fact that, as Claire Préaux has argued, the economy collapsed in Egypt during this period.[17] Nevertheless, the papyri show that the tonnage increased considerably. This rise cannot be observed via the shipwrecks, because shipwrecks were a rarity in the Eastern Mediterranean.

It is more surprising that only a few shipwrecks are known to have occurred in the Western Mediterranean during the first half of the second century, because there were plenty of them during the following centuries. Could the lack of shipwrecks in Western waters during this period mean that shipbuilding might have been delayed in the Eastern area? Could Eastern fleets have included bigger ships than Roman ones? Could there have been a difference in the tonnage of ships between the two parts of the Mediterranean?

Actually, the lack of shipwrecks in the Western Mediterranean in the first half of the second century should not be exaggerated (Table 5.2). It seems that the population of Rome reached nearly 200,000 inhabitants

---

16 About these ships, which were carrying 3000 *medimnoi* (about 90–117 tonnes), see the honorific decrees studied by Lionel Casson: Casson 1956–1957; 1971, 183–84. For a discussion about these inscriptions, see Nantet 2016, 116–17.
17 Préaux 1939, 137.

in the beginning of the third century, and twice more one-and-a-half centuries later. These rough estimations should be considered very cautiously — but no doubt the grain supply required by such a population came to several hundred shipments per year.

Table 5.2 Estimated number of shipments required for the supply of Rome.

Date	Population (estimated by P.A. Brunt[18])	Yearly consumption (in modii)[19]	Quantity of imported wheat, including the 20% damaged during the transportation[20]	Equivalent number of shipments of 30,000 modii [21]
270	180,000	7,560,000	9,072,000	302
130	375,000	15,750,000	18,900,000	630

The importance of the grain supply is confirmed by the many gifts offered to the Romans by the Hellenistic kings of the Western Mediterranean, such as Hiero II and Massinissa (Table 5.3). These gifts must also have required a high number of large ships. Thus, the evidence shows that the increase in tonnage was not restricted to the Eastern Mediterranean. Many large ships were sailing on the Western waters too. Thus, there was an overall growth in tonnage throughout the Mediterranean.

That initial overall rise indicates the importation of wheat and stone above all. An enormous volume of wheat would have been shipped in order to supply the cities, like Rome or Alexandria, which were becoming bigger and more populous. Stone, like wheat, also required big ships. During the Hellenistic era, cities built porticoes and fortifications, which both demanded large amounts of stone. Both wheat and stone were shipped in huge quantities.

As in former centuries, wine was still a very sought-after commodity. But cargoes of wine were usually very small, no more than a few dozen tonnes. Other kinds of merchandise, such as copper, oil, and even wool, were also shipped in limited quantities and were therefore not conveyed in large vessels.

---

18 Brunt 1971, 69.
19 A. Tchernia considers an annual consumption of 42 modii, i.e., 286 kg, for one person. Tchernia 2000.
20 Tchernia 2000.
21 The amount of 30,000 modii (204 tonnes) has been chosen, because it is very close to the amount of 10,000 artabae (227 tonnes), which was probably the common tonnage during this period.

Table 5.3 The gifts of the Western Mediterranean powers to the Romans from the second half of the 3rd century to the first half of the 2nd century BCE (after Garnsey 1996, 241–246).

Date	Event	Amount (in modii)	Equivalent number of shipments of 30,000 modii[22]
237	Gift from Hiero II (Eutropius, 3, 1)	200,000	7
215	Gift from Hiero II to the Roman army protecting the Adriatic sea (Livy, 23, 38, 13)	300,000	10
200	Gift from Massinissa to the Roman army in Macedonia (Livy, 31, 19, 4)	400,000	13
198	Gift from Massinissa to the Roman army in Greece (Livy, 32, 27, 2)	200,000	7
191	Gift from Massinissa (Livy, 36, 4, 8)	750,000	25
191	Gift from the Carthaginenses (Livy, 36, 4, 8)	1,000,000	33
170	Gift from the Carthaginenses (Livy 43, 6, 11)	1,500,000	50
170	Gift from Massinissa (Livy, 6, 13-14)	1,500,000	50

Thus, these big ships were built essentially to carry grain and stone, heavy goods whose value was lower than wine. That first rise in tonnage was not caused by a desire to make money. Instead, it was connected to efforts to gain more prestige, as well as to political and military issues. It should be noted that this growth only happened when the three main Hellenistic kingdoms declined, and *before* Roman rule.

5.3. A Second Rise from the End of the Second Century to the Beginning of the First Century?

A second increase occurred from the end of the second century to the beginning of the first century BCE, and this rise was confined to

---
22 See Table 5.2, note 21.

the Western Mediterranean. Indeed, the Alberga (500–600 tonnes),[23] the Madrague de Giens (402.5 tonnes)[24] and Mahdia (230–250 tonnes)[25] shipwrecks are all large ships. Their tonnages were far beyond those of former centuries, and they were evidently owned by the biggest merchantmen of their age.[26]

However, we lack papyrological evidence for this period.[27] It would be tempting to conclude that tonnage dropped suddenly in Egypt during the first century. But the lack of papyri does not mean that there were no longer any ships on the Nile. It is possible that no papyri originating in this period were found because of the timing of the discoveries. However, it is notable that the vast majority of the edited papyri date from the Late Hellenistic or the Roman era. So how could the lack of papyri from the first century BCE be explained? The transportation of the wheat may have been organized differently during this time, although the only papyrus from this period in our list, the *SB 5 8754*, does not seem to show this. It could also be that this lack of papyri might have been the result of a change in the Ptolemaic administration, which managed the grain supply. The Ptolemaic authorities might have systematically ceased producing these documents for some unknown reason. So far, we have only conjecture, and no definite answers have been established.

Nevertheless, it should be noted that absolutely no sources in the Eastern Mediterranean show an increase in tonnage, whether they be literary, epigraphical, or even archaeological. The Doric capital and the eight column drums of the Kızılburun shipwreck did not weigh much more than 50 tonnes.[28] However, this does not necessarily mean that there were no large ships sailing in the Eastern Mediterranean. It seems that the biggest ships sailed on only a few key routes, such as that from Greece to the Western Mediterranean, in order to convey works of art (including marble stones). For instance, the cargoes of the Mahdia or Antikythera shipwrecks were among the biggest of that period.

---

23 The cargo has been estimated at between 11,000 and 13,500 amphorae, i.e. 500–600 tonnes, by Pomey and Tchernia 1978.
24 Tchernia et al. 1978; Pomey 1982.
25 Hellenkemper Salies 1994.
26 Pomey and Tchernia 1978.
27 Hauben's most recent list, which is not focused on ships but on owners, shows the same lack of papyri for the first century.
28 Carslon 2010.

The other main route was from Italy to Gaule in order to sell wine. Strabo tells us that a slave could be bought only for an amphora.[29] Many ships subsequently wrecked, especially the Madrague de Giens, were involved in that trade. In fact, even the smallest ships carried wine. The vessel that sank off Cap Bénat carried no more than three *pithoi*, i.e. 3.3 tonnes,[30] but the wine trade in that area was so successful that Roman merchants used large vessels, called *dolia*, to carry larger containers than amphorae.[31]

Although the data are patchy, there might have been more of these routes. For instance, the hulls found in Caesarea[32] and Antirhodos[33] almost certainly belonged to medium-sized or, perhaps, even large ships. They can likely be dated to the first century CE. However, we cannot totally rule out the possibility that both of these shipwrecks were from earlier or later. No definite date can be asserted as long as no dendrochronological analysis has been published.

Contrary to the first rise in tonnage that we have considered in this chapter, the second increase did not concern only the transportation of cargoes such as grain or stone, but also wine or works of art, whose value was much higher. This increase was made possible by the considerable wealth of the Roman elites (and maybe some vassal princes of Rome) — the routes related to this second rise were all connected to the city. Such evidence reveals the significant changes to maritime trade caused by Roman rule.

5.4. The Common Reasons for the Two Increases

Thus these two changes were caused by different factors, but both originally had the same roots. Above all, new techniques allowed the construction of bigger ships,[34] even giant ones, such as the *Syracusia*.[35] Nonetheless, one of the major obstacles to the increase in tonnage was the lack of deep harbours. The *Syracusia*, which could load up to 2580

---

29 Diodorus Siculus, 5.26.3.
30 Joncheray 1997.
31 Marlier 2008; Heslin 2011.
32 Fitzgerald 1994, 1995.
33 Sandrin et al., 2011.
34 See the contribution by Pomey in chapter 3.
35 Athenaeus, 5.206^d–209^b.

or 2706 tonnes,³⁶ could not enter many harbours because her draft was too large. Thus, she was useless in this capacity and Hiero II gave her to Ptolemy III. This demonstrates how restrictive the harbour depth could be for boats like these.

The authorities undertook to dredge their harbours in order to make them as deep as they could. During the digging of the tube station *Piazza Municipio* in Naples, an excavation led to the discovery of an ancient harbour, which included several shipwrecks, including Napoli A, B and C. But there were some strange marks on the bottom, as if the harbour base had been scratched. These marks were made by a dredger between the fourth and second centuries BCE.³⁷

It was not easy for the authorities to reserve the deepest parts of the harbour for the bigger ships. The regulations in Thasos reveal that the many small ships were cluttering the harbour and that they were docked in the deepest areas, which were the only places the bigger ships could dock. Actually, the first harbour regulations seem to be linked to the need to preserve the depth of the harbour for the bigger ships.

The technical developments of both ship and harbour technologies allowed the increase in tonnage. They made the growth *possible*. But the root of the increase is located in the development of cities in the Hellenistic world. Those cities became more populated, and more people meant more producers, more consumers, and more trade. This trade required more ships to supply these cities with wheat and stone. Not only did the population grow, but grain and stone had to be transported across *longer* distances. Trade was no longer limited to the Aegean Sea or to any other part of the Eastern Mediterranean. From then on, certain ships sailed through many seas in the whole Eastern Mediterranean, and sometimes beyond, as shown by the Mahdia and Antikythera shipwrecks. In other words, the organization of trade on a much larger scale than before led to the rise in tonnage.

---

36 For more information about the estimation of her tonnage, see Nantet 2016, 126–31; Nantet forthcoming.
37 Giampaola et al. 2004, 2005. For more information about harbour maintenance, see Nantet 2016, 223–28.

5.5. Conclusion

There was a notable increase in tonnage throughout the Mediterranean over a period of several centuries. At the beginning of the Late Hellenistic period, that increase occurred across the entire Mediterranean. From the end of the second century to the first century BCE, there was a second rise, which was restricted to specific parts of the Mediterranean.

The situation changed in the Imperial period. Even though it seems that large ships may have continued to carry large amounts of wine over the seas during the first century CE, they may have been less numerous in the following centuries. On the contrary, wheat and stone cargoes were conveyed by ships that became larger and larger. Indeed, the supply of grain to Rome became a major issue for the emperors who wished to watch over the situation in the streets of the capital city. Moreover, they wanted to provide Rome with the most impressive monuments and therefore required large quantities of marble.

Bibliography

Primary Sources

Literary Sources

Bird, H. W., ed. 1993. *Eutropius: The Breviarum ab urbe condita of Eutropius: The Right Honourable Secretary of State for General Petitions, Dedicated to Lord Valens, Gothicus Maximus & Perpetual Emperor*. Liverpool: Liverpool University Press.

Gulick, C. B., ed. 1957. *Athenaeus. The Deipnosophists*. Cambridge, MA: Loeb Classical Library.

Moore, F. G., ed. 1951. *Livy. History of Rome*. Cambridge, MA: Loeb Classical Library.

Oldfather, C. H., ed. 1962. *Diodorus of Sicily*. Cambridge, MA: Loeb Classical Library.

Inscriptions

Launey, M. 1933. 'Inscriptions de Thasos.' *Bulletin de correspondance hellénique* 57:394–415.

Papyri

Hauben, H. 1978. 'Nouvelles remarques sur les naucléres d'Egypte à l'époque des Lagides.' *Zeitschrift für Papyrologie und Epigraphik* 28:99–107.

Hauben, H. 1997. 'Liste des propriétaires de navires privés engagés dans le transport de blé d'état à l'époque ptolémaïque.' *Archiv für Papyrusforschung und verwandte Gebiete* 43:31–68.

Meyer-Termeer, A. J. M. 1978. *Die Haftung der Schiffer im griechischen und römischen Recht*. Zutphen: Terra.

Archaeological Excavations

Carlson, D. N. and W. Aylward. 2010. 'The Kızılburun Shipwreck and the Temple of Apollo at Claros.' *American Journal of Archaeology* 114:145–59. https://doi.org/10.3764/aja.114.1.145

Demesticha, S. 2011. 'The 4th-Century-BC Mazotos Shipwreck, Cyprus.' *IJNA* 40:39–59. https://doi.org/10.1111/j.1095-9270.2010.00269.x

Fitzgerald, M. A. 1994. 'Chapter VI: The Ship.' In *The Harbours of Caesarea Maritima. Results of the Caesarea Ancient Harbour Excavation Project (1980–85) vol. 2*, edited by J. P. Oleson, 163–223. Oxford, Bar International Series 594.

Fitzgerald, M. A. 1995. 'A Roman Wreck at Caesarea Maritima, Israel: A Comparative Study of its Hull and Equipment.' PhD diss., Texas A&M University.

Giampaola, D., V. Carsana, and G. Boetto. 2004. 'Il mare torna a bagnare Neapolis. Parte II: dalla scoperta del porto al recupero dei relitti.' *L'Archeologo Subacqueo* 10: 3:15–19.

Giampaola, D., and V. Carsana. 2005. 'Neapolis. Le nuove scoperte: la città, il porto e le macchine.' In *Eureka! il genio degli antichi, catalogo della mostra, Museo Archeologico Nazionale di Napoli, 11 luglio 2005–9 gennaio 2006*, edited by E. Lo Sardo, 116–22. Naples: Electa Napoli.

Hellenkemper Salies, G., ed. 1994. *Das Wrack: der antike Schiffsfund von Mahdia*. 2 vols. Cologne: Rheinland-Verl.

Joncheray, J.-P. 1997. 'Bénat 2, une épave à dolia du Ier s. av. J.-C.' *CAS* 13:97–119.

Pomey, P. 1982. 'Le navire romain de La Madrague de Giens.' *Comptes rendus des séances de l'Académie des inscriptions et belles-lettres (Paris)* 126 (1):133–55.

Sandrin, P., A. Belov, and D. Fabre. 2013. 'The Roman Shipwreck of Antirhodos Island in the *Portus Magnus* of Alexandria, Egypt.' *IJNA* 42:44–59. https://doi.org/10.1111/j.1095-9270.2012.00363.x

Steffy, J. R. 1985. 'The Kyrenia Ship: An Interim Report on its Hull Construction.' *American Journal of Archaeology* 89:71–101.

Tchernia, A., P. Pomey, and A. Hesnard. 1978. 'L'épave romaine de la Madrague de Giens (Var).' *Gallia* Suppl. 34. Paris: Éditions du Centre National de la Recherche Scientifique.

Secondary Sources

Arnaud P. 2015. 'La batellerie de fret nilotique d'après la documentation papyrologique (300 avant J.-C.–400 après J.-C.).' In *La Batellerie égyptienne, Archéologie, histoire, ethnographie*, edited by P. Pomey, 99–150. Études Alexandrines 34. Alexandrie: Centre d'Études Alexandrines.

Casson, L. 1956. 'The Size of Ancient Merchant Ships.' In *Studi Aristide Calderini e Roberto Paribeni*, vol. 1. Milan: Ceschina, 231–38.

Casson, L. 1971. *Ships and Seamanship in the Ancient World*. 2nd ed. Ann Arbor, MI: Princeton University Press.

Christopoulou, A., A. Gaolou, and P. Bougia, eds. 2012. *The Antikythera Shipwreck: The Technology of the Ship, the Cargo, the Mechanism*. Athens: National Archaeological Museum.

Garnsey, P. 2009 [1988]. *Famine and Food Supply in the Graeco-Roman World: Responses to Risk and Crisis*. Cambridge: Cambridge University Press. https://doi.org/10.1017/CBO9780511583827

Heslin, K. 2011. 'Dolia Shipwrecks and the Roman Wine Trade.' In *Maritime Archaeology and Ancient Trade in the Mediterranean*, edited by D. Robinson and A. Wilson, 157–68. Oxford Centre for Maritime Archaeology monograph 6. Oxford: Oxford Centre for Maritime Archaeology, Institute of Archaeology.

Houston, G. W. 1988. 'Ports in Perspective: Some Comparative Materials on Roman Merchant Ships and Ports.' *American Journal of Archaeology* 92:553–64.

Kaltsás, N. E., E. Vlachogianni, and P. Bougia, ed. 2012. *The Antikythera Shipwreck: the Ship, the Treasures, the Mechanism: National Archaeological Museum*, April 2012–April 2013. Athens: Kapon Editions.

Marlier, S. 2008. 'Architecture et espace de navigation des navires à dolia'. *Archaeonautica* 15:153–73. https://doi.org/10.3406/nauti.2008.920

Nantet, E. 2016. *Phortia. Le tonnage des navires de commerce en Méditerranée du VIIIe siècle av. l'è. chr. au VIIe siècle de l'è. chr*. Rennes: Presses Universitaires de Rennes.

Nantet, E. 2017. 'The accuracy of the tonnage formula, and the correcting coefficient.' In *Baltic and Beyond: Change and Continuity in Shipbuilding: Proceedings of the Fourteenth International Symposium on Boat and Ship Archaeology (Gdansk 2015: ISBSA, 14)*, edited by J. Litwin and W. Ossowski,

165–170. Gdańsk: National Maritime Museum. http://opac.regesta-imperii. de/id/2427830

Nantet, E. Forthcoming. 'The Tonnage of the Syracusia: a metrological reconsideration.' In *'Under the Mediterranean': The Honor Frost Foundation Conference on Mediterranean Maritime Archaeology (20th – 23rd October 2017), Short Report Series.*

Pomey, P. and É. Rieth. 2005. *L'Archéologie navale*. Paris: Errance.

Pomey, P. and A. Tchernia. 1978. 'Le tonnage maximum des navires de commerce romains.' *Archaeonautica* 2:233–251.

Tchernia, A. 2000. 'Subsistances à Rome: problèmes de quantification.' In *Mégapoles méditerranéennes: géographie urbaine rétrospective: actes du colloque, Rome, 8–11 mai 1996,* edited by C. Nicolet, 751–60. Paris: Maisonneuve et Larose.

Thompson, D. J. 1983. 'Nile Grain Transport Under the Ptolemies.' In *Trade in the Ancient Economy,* edited by P. Garnsey, K. Hopkins, and C. R. Whittaker, 64–75. London: Chatto and Windus.

Weinberg, G. D., V. R. Grace, G. R. Edwards, and H. S. Robinson. 1965. *The Antikythera Shipwreck Reconsidered*. Philadelphia: American Philosophical Society.

6. A Note on the Navigation Space of the *Baris*-Type Ships from Thonis-Heracleion

Alexander Belov

The available data on local boat-building techniques during the Late (664–332 BC) and Ptolemaic Periods (332–30 BC) of Ancient Egypt received a considerable boost from the more than sixty Ancient Egyptian ships that were found on the site of Thonis-Heracleion in 2000. Many of these ships seem to belong to the *baris*-type as described in Herodotus in his *Historia*. This chapter is an attempt to determine the space of navigation of these ships by examining the direct evidence derived from their construction, as well as indirect evidence drawn from the state of the ships' timbers and the results of reconstruction of their hulls, and of their propulsion and steering systems.

The site of Thonis-Heracleion is situated in the Bay of Abukir to the west of Alexandria, and it has been undergoing excavations by the European Institute for Underwater Archaeology (IEASM) since 1999.[1] The city had a rather complicated topography that abounded with peninsulas, canals and semi-enclosed areas of water. The passages between the sand dunes connected the coastal lagoon and the harbours of Heracleion with the Canopic branch of the Nile (Fig. 6.1).[2] The geographical situation of the

---
1   Goddio 2007, 102–14.
2   For the latest information on Heracleion's topography see Goddio 2011; Fabre et al. 2013; Goddio et al. 2015; Goddio 2015.

city corresponds fairly well with the concept of a maritime gateway.³ The city served as customs station for foreign ships going up the Nile and it was occupied already in the late eighth or early seventh century BC, though the second century BC was its golden age.⁴ In the Late Period the city controlled access to the Canopic branch of the Nile and was engaged in trade with Greece.⁵

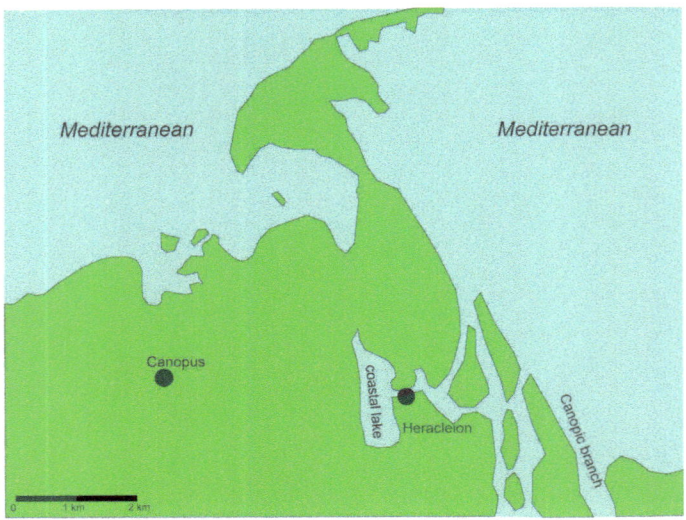

Figure 6.1 Simplified topography of the Canopic region
(After Goddio 2007, 17, fig. 1.15.)

To this day, sixty-four shipwrecks have been discovered on the site of Thonis-Heracleion.⁶ Although only preliminary studies were carried out on the majority, several ships have been excavated.⁷ Numerous original features shared by many of these ships⁸ seem to bear witness to an

---

3   Term first proposed by Hirth (1978).
4   Yoyotte 2001; Fabre 2008.
5   Robinson and Goddio 2015.
6   The actual number of ships probably exceeds one hundred. During the survey with a high-tech sub-bottom profiler in autumn 2016, many dozens more were discovered, some of them five meters under the clay (F. Goddio, personal communication). These ships are remarkably well preserved. For the origins of this vast assemblage of ancient vessels, see Robinson 2018.
7   Ships numbered 17, 43, 61 and 11.
8   Preliminary studies show that probably ships numbered 3, 8, 10, 17, 23, 43, 44, 45, 50, 51 and 63 belong to a *baris* type. In this author's opinion, that may well be the case for the majority of ships preserved on the site of Thonis-Heracleion.

archaeologically unattested constructional type, which finds parallels in Herodotus' description of a freighter (barge) called the *baris* (*Historiae* 2.96, c. 450 BC).[9]

6.1. Main Characteristics of the *Baris* as per Herodotus and New Archaeological Data

The Greek term *baris* (βᾶρις) probably originates in the Ancient Egyptian boat type called *br* (*byr*)[10] that first appears in the Eighteenth Dynasty and refers to a sea-going craft.[11] Demotic documents mentioning *br* (*byr*) are not numerous and contrary to hieroglyphic texts most of them probably refer to Nilotic cargo boats.[12] Textual evidence from Greek papyri suggests that the *baris* was primarily a multipurpose freighter and transport vessel.[13] Gradually replaced by other types, first of all probably by the *kerkouros*, the *baris* is last mentioned in the papyrus dated to 125 BC.[14]

The excavations of Ship 17 from Thonis-Heracleion helped to clarify several references from Herodotus' description that had previously been incomprehensible. Thus, the main features of the construction of the *baris* may be summarized in the following terms. The *baris* was a flat-bottomed freighter built from local acacias. A central keel-plank or a kind of proto-keel (Ship 17)[15] did not project beneath the crescent-shaped hull. The planking of this ship consisted of short planks arranged like 'courses of bricks'. Long tenons reaching 2 m in length passed inside

---

9 Belov 2014, 2015b, 2019.
10 Casson 1971, 341, note 64.
11 Ibid.; Vinson 1994, 44–5; 1998, 252.
12 Ibid., 252–53.
13 Casson 1971, 340, note 60; 341, note 64; Vinson 1998, 254. Vinson cites two documents that might indicate a military use for *br* ships (Vinson 1998, 253). According to line 12 of the Rosetta Stone (196 BC) the defensive fleet of Ptolemy V contained νῆες (ships) in the Greek text that correspond to the hieroglyphic *kbn.wt* and to the Demotic *byry*. Darnell (1992, 72–73, notes 21 and 54) suggests a parallel between these ships and those employed by Rameses III (1184–1153 BC) to defend the Delta against the Sea Peoples. Another example is the Roman *P. Krall* 14/8 mentioning *br* ships as part of a naval fleet. In papyrus *W.Chr.* 11 A (123 BC) a *baris* transports soldiers (See Arnaud 2015b, 116). The employment of freighters during a war for purposes such as transporting the troops or as auxiliary fighting units is quite obvious and does not require further comment.
14 Arnaud 2015b, 116.
15 Belov 2015a.

rectangular channels that were cut in the middle of the planks' edges, and were pegged to the planking at the extremities. At the same time, the tenons wedged the through-beams to the planking. The inner joints between the planks were sealed with papyrus. The boat was steered with an axial rudder that passed through an opening in the keel. The *baris* was a sailing ship, but according to Herodotus, it could only travel upstream with the help of a fresh breeze. Herodotus did not mention oars, and no traces of rowing arrangements were found on ships of this type from Heracleion. According to Herodotus, *barides* were built with quite a different carrying capacity and these ships were numerous on the Nile. Ship 17 would have been about 27–28 m long with a beam of 8 m that gives a length-to-width ratio of around 3:4. The ship had a displacement of about 150 metric tonnes, a draft of 1.6 m and a tonnage of approximately 112 metric tonnes.

6.2. Navigation Area of the *Baris*-Type Ships

6.2.1. Written Sources

As mentioned above, it seems that the term *baris* radically changed its meaning from the New Kingdom to the early Ptolemaic period, when, according to available documentation, the ship was primarily employed on the river. Thus, here again, the text of Herodotus, contemporary with the *baris*-type ships from Heracleion, appears to be the most important source for the current discussion.

Herodotus' description of the *baris* comes sequentially after information on different aspects of life in the Delta, and it is logically linked to the description of Delta shipping in fragment 2.179.[16] These observations give more weight to the arguments for the Delta origins of the *baris*, rather than an origin in the Nile valley.

According to Herodotus, the *baris* under sail could overpower the Nile's current only in case of a strong wind; otherwise, she was hauled from the bank.[17] Herodotus also describes the original technique used by the Egyptians for steering the *baris* downstream with a help of a small raft

---

16 Vinson 1998, 252.
17 Arnaud (2015a, 109) judiciously remarks that hauling is possible from a firm bank only, something difficult to achieve on a river with an ever-changing hydrological regime.

and an anchor, their joint action straightening her course.[18] Apparently the hauling of a ship upstream[19] and a sophisticated technique for the descent both favoured the vessel's employment on the river.

Book two of Herodotus contains another very important fragment related to this topic:

> Now in old times Naucratis alone was an open trading- place, and no other place in Egypt: and if any one came to any other of the Nile mouths, he was compelled to swear that he came not thither of his own will, and when he had thus sworn his innocence he had to sail with his ship to the Canobic mouth, or if it were not possible to sail by reason of contrary winds, then he had to carry his cargo round the head of the Delta in boats ['*baris*' in the original text - AB] to Naucratis: thus highly was Naucratis privileged.[20]

It is significant that Herodotus used different terms for the foreign seagoing vessel (ναῦς) and for the ships employed for local transportation (*baris*). This testimony confirms that the *barides* could operate beyond the Delta and thus belonged to a class of fluvic-maritime vessels. The following sections inquire whether this conclusion can be applied to the *barides* from Thonis-Heracleion.

6.2.2. Context of Ships from Thonis-Heracleion[21]

It is important to underline the fact that the *baris* ships were quite numerous at Heracleion. It is still difficult to determine with precision the depth of the port facilities, but the coastal lagoons are quite shallow

---

18 *Historiae* 2.96. A physical model developed during an interesting experiment carried out by Goyon in collaboration with the Central Hydraulic Laboratory of France proved the efficiency of the technique described by Herodotus. See Goyon 1971, 38–41, annex 1. Mathematical manipulations proposed in a subsequent publication by Wehausen et al. (1988) fully confirm the results of the modelling.

19 Cf. Casson 1965.

20 *Historiae* 2.179. Trans. Macaulay 1890. 'ἦν δὲ τὸ παλαιὸν μούνη Ναύκρατις ἐμπόριον καὶ ἄλλο οὐδὲν Αἰγύπτου· εἰ δέ τις ἐς τῶν τι ἄλλο στομάτων τοῦ Νείλου ἀπίκοιτο, χρῆν ὀμόσαι μὴ μὲν ἑκόντα ἐλθεῖν, ἀπομόσαντα δὲ τῇ νηὶ αὐτῇ πλέειν ἐς τὸ Κανωβικόν· ἢ εἰ μή γε οἷά τε εἴη πρὸς ἀνέμους ἀντίους πλέειν, τὰ φορτία ἔδεε περιάγειν ἐν βάρισι περὶ τὸ Δέλτα, μέχρι οὗ ἀπίκοιτο ἐς Ναύκρατιν. οὕτω μὲν δὴ Ναύκρατις ἐτετίμητο.' Herodotus does not mention Thonis. The hypothesis dealing with this omission can be found in Höckmann 2008–2009, 115, 124.

21 Different hypotheses regarding the origin of ships' accumulations (land reclamation or blockship barrier) may be found in Robinson 2015. Cultural, socio-economic and geopolitical contexts are considered in Fabre 2015.

and usually about two to three meters deep. According to the recent Sediment Profile Imaging (SSPI) survey, the maximum depth in the ports of Heracleion did not exceed 4.5–5 m.[22] Thus these ships with obviously shallow drafts were quite adapted to this environment. In addition, navigation on the Nile was highly seasonal[23] and the smaller specimens of the *barides* seem to have been advantageous, as they could operate for a longer time than other types.[24]

An interesting clue is offered by the anchors that were found in great numbers (more than 700) in the harbours of Heracleion. The anchors appear in different types but most of them are triangular stone composite anchors with two round front openings for wooden arms and one transverse opening for the cable.[25] The majority of anchors are about 70–90 cm long and approaching a hundredweight. Some of these anchors were found on board the *barides* in a position relevant for mooring. This is certainly true in the case of Ship 43, which had a 100 kg anchor placed in vertical position at the bow.[26] These anchors were probably handled with a help of tackles and a mast-derrick.[27]

While there is a plenty of archaeological evidence of Ancient Egyptian anchors on board sea-going vessels,[28] these were of no real use on the Nile.[29] Instead, a wooden stake[30] was driven into the muddy shore with

---

22   Cataudella et al. 2015, 73, Table II. F. Goddio, personal communication.
23   Cooper 2011, 195; 2012, 61; 2012a.
24   See Robinson 2015, 213. Cooper (2012a, 26) cites the nineteenth-century sources according to which the ships with deadweight of 60 tonnes were not able to navigate in the Delta during five months of the year.
25   See Nibbi 1991.
26   Calibrated date ^{14}C for wooden arms: 405 cal–208 cal BC. Dimensions: 75 x 50 x 18 cm.
27   Basch 1987, 66–67; Frost 1995. The destination of numerous huge anchors found in Heracleion could have been different. For example, the largest among two anchors found at the bows of Ship 51 (?) weights 630 kg (!) (calibrated date ^{14}C for wooden arms: 396 cal–198 cal BC. Dimensions: Anchor 1 — 106 x 80 x 26 cm; Anchor 2 — 154 x 94 x 30 cm). The hull of this ship was only partially preserved but several parameters of its planking indicate that the length of the ship did not exceed 20–25 m, so it was not of extraordinary size. Thus, it might have been that the largest anchors of Heracleion were used as mooring anchors. This idea was introduced by several members of the jury for my PhD thesis on 31 January 2014 (P. Arnaud, P. Pomey, F. Goddio). The same hypothesis had already been proposed for the pyramidal stone anchors from Zea (see Tzalas 1999). Bronze Age stone anchors weighing 850 and even 1350 kg were found in Kition (see Frost 1985).
28   Basch 1985, 1994; Zazzaro 2007, 2011; Zazzaro and Abd el-Maguid 2012; Tallet 2013, 2015.
29   Basch 1985, 457; 1994.
30   *mnit* or *n^{c}yt* (Jones 1988, 198, n. 4, 199, n. 8).

a mallet,[31] as shown by iconographical sources.[32] Furthermore, there is no solid iconographical proof for the use of anchors on the Nilotic ships. However, it might have been that mooring techniques changed during the Late Period following the increase of maritime trade and fluvio-maritime traffic, for which the anchors were absolutely indispensable. Several 'elongated composite stone anchors'[33] were discovered in a riverine environment in Egypt.[34]

The Delta was a very particular region between the river and the sea, characterized by its unique navigational conditions and hazards[35] caused by the varying geomorphology, geology, hydrology, and meteorology of this area. According to Yoyotte, the ancient name of the city — Thonis (Θῶνις in Ancient Greek sources) — originates in the indigenous name of the coastal lagoon (*henet/hone*) that existed there in Antiquity.[36] The water of the lagoon was only slightly brackish and this is confirmed by numerous finds of the bones of Nile catfish (*Siluriformes*) and other fresh-water organisms.[37] The sedimentology of a coastal lagoon is very different to that of a river and includes sediments ranging from coarse sand to silt and clay.[38] Many hundreds of discoveries from Heracleion prove that this environment allowed regular employment of marine-type anchors. This is not surprising, taking into consideration the intense shipping and manoeuvring in the restricted harbour space, and the limited total length of wharfs. It seems that in the Ptolemaic period, the river was perceived as an extension of the sea.[39] The structure of river administration followed the maritime model, as did mooring procedures in a *hormos*. These

---

31 *ḥrpw* (Jones 1988, 201, n. 12).
32 For more on the Ancient Egyptian iconography of mooring, see Doyle 1998, 220–35.
33 Frost 1970, 381.
34 Abd el-Maguid 2015.
35 Cooper 2012, 2012a.
36 *henet/hone* → T(hone) → Thonis. See Yoyotte 2001; 2013, 298–9, 307–8, 349–52. This specific geography is described in classical sources: Heliodorus, *Aethiopica* 5; Achilles Tatius, *Leucippe and Clitophon* 4.12.7–8; Diodorus Siculus, *Bibliotheca Historica* 1.31.2–5. For discussion see Fabre 2015, 180–84.
37 Goddio 2007, 111.
38 For more information about the geomorphology of coastal lagoons see Bird 1994. According to El-Wakeel and Wahby (1970) 'the predominant type of sediment covering the bottom of the lake [Manzalah — A.B.] is the complex type sand-silt-clay followed in abundance by the clayey sand and silty clay. There is a basinward increase in grain size of sediments.'
39 Arnaud 2015b, 104–05.

factors dictated the choice of the Mediterranean style of mooring[40] and the employment of the marine variety of anchors that were also necessary for open mooring.[41]

The acacia wood used as raw material, and many other features of indigenous shipbuilding, correspond well with etymological arguments and written sources testifying that the *baris* was undoubtedly a local type. These ships used anchors of a marine type as confirmed by Ship 43. However, this fact is not decisive as the anchors could have been used for mooring beyond the sandbar separating the estuary from the open sea, or only within its limits and in the harbours of Heracleion. In order to determine the navigational area of these ships, it is necessary to look closer at their construction.

6.2.3. Direct Evidence from Ships' Construction

Many features of the *baris*-type ships indicate their river origins. The hull of the *baris* was constructed with very short planks. In the case of Ship 17, the average length of the planks was only 192 cm, while the segments of the proto-keel did not exceed 3 m in length.[42] The proto-keel did not protrude, and that was an advantageous option for river navigation. The most ancient Egyptian term we know of referring to a keel or to a keel plank (*pipit*[43]) may mean a 'mud-kneader'.[44] The same type of flat keel has been incorporated into the construction of the modern *nuggars* of the Upper Nile.[45]

All the elements of the ships' inner structure were characterized by a strong asymmetry and a roughness of execution. Thus, usually the beams were not horizontal, and were made of irregularly-shaped branches. All Ancient Egyptian sea-going vessels known from texts and from the archaeological record were built of imported wood, while the *barides* of Heracleion were built from local species of acacia which had

---

40   Also known today as 'med mooring' or 'Tahitian mooring', this technique means that the vessel sets a temporary anchor off the pier and then approaches it at a perpendicular angle. The vessel then runs two lines to the pier.
41   Ibid., 104.
42   Belov 2014.
43   Jones 1988, 164, n. 52.
44   Goedicke 1975, 95; Janssen 1975, 379.
45   Clark 1920, 49; Hornell 1943, 28.

been employed to construct river-faring boats since the Old Kingdom (2686–2160 BC).[46]

The joints of the planking of these ships were different from the double rows of the relatively small tenons and lashings mainly associated with the planking of Ancient Egyptian sea-going ships.[47] Moreover, the tenons of the *baris* were pegged and that was never the case with the planking of sea-going ships, which employed free tenons to facilitate the assembly and disassembly of their hulls for transportation and storage.[48] The vessels of Heracleion which have been studied so far were undecked.

Thus, the constructional features of the *barides* seem to indicate that these ships were not really adapted for conditions on the open seas.

6.2.4. Reconstruction of the Hull: Supplementary Data

The preliminary reconstruction of Ship 17[49] suggests a crescent-shaped hull with considerable overhangs at both extremities.[50] The overall length of the ship should have been about 27–28 m with a beam of 8 m. Its displacement was close to 150 tonnes, with a tonnage of about 113 tonnes.[51] This was one of the largest *barides* known in Heracleion.

6.2.5. Longitudinal Structure

The short segments of the planking presented serious challenges for the longitudinal structure of the *baris*-type ships, as it seems that about

---

46 Nilotic freighters *sekhet* and *satch* built by general Weni (*Wnj*) during the rule of Pepi I (Sixth Dynasty, 2345–2181 BC). According to the text they were 32 m long. Together with tamarisk, acacia wood was identified as the construction material of the freighter boats from Lisht (Middle Kingdom, c. 1950 BC). See Ward 2004, 15. Traditional boats of the Upper Nile are still built of *Acacia nilotica* (See note 29). Acacia also dominates as the constructional material for the ships of Thonis-Heracleion. Preliminary xylological analysis showed that, among 63 shipwrecks, about 80% have planking made of acacia. See Fabre and Belov 2012, 109–10. Ship 17 of Thonis-Heracleion was entirely built of acacia. See Belov 2014.
47 Timbers from Mersa Gawasis and Ayn Sukhna. See Ward and Zazzaro 2010; Pomey 2012a, 2012b. However, double rows of mortises were equally attested in planks from Lisht that belonged to a river-going freighter. See Haldane 1993, 237.
48 See Ward 2007; Pomey 2012a, 2012b.
49 Belov 2019, chapter 3.1.
50 An iconographic parallel is provided by one of the ships depicted on the mosaic of Palestrina (c. 125 BC), which was recently identified as a *baris* by Pomey (2015, 164–66).
51 Belov 2015a, 206–07.

only three-fifths of the overall length of their crescent-shaped hulls was supported by the water.[52] It is not yet clear how this problem was solved, although a bulwark might have played an important role in the longitudinal structure of Ancient Egyptian boats, to counterbalance the hogging of the hull.[53] No other means for maintaining the longitudinal strength of the hull have been discovered in the construction of the *barides*.[54] Hypothetically, the through-beams were capable of significantly reinforcing their longitudinal structure.[55]

6.2.6. Shallow Draft

Ship 17, being one of the largest specimens of the *baris* in Heracleion, would have a shallow draft of about 1.6 m that would have been a definite advantage for navigation on the river and within the shallow lagoons of the Delta, like those of Heracleion. The depth at the mouths of the Nile must have been inconsiderable too.[56]

An interesting parallel is suggested by the Arab fishing boats of the Manzala[57] and Borollos[58] lakes. These boats, with a shallow draft, are perfectly adapted to traditional fishing inside a coastal lagoon.[59]

---

52 According to preliminary results of the modelling, this ratio was about 66% in the case of Ship 17.
53 Haldane 1993, 234–35; Vinson 1997.
54 A hogging truss was used in the construction of Egyptian sea-going ships at least until the New Kingdom (exemplified by the sea-going vessels of the Punt expedition launched by Queen Hatshepsut, Eighteenth Dynasty, 1473–1458 BC). However, there is no evidence that a hogging truss was employed on ships from Heracleion. On the other hand it remains a possibility that a rope truss was employed during the *constructional* phase to pre-stress the hull against the hogging. Although this element disappears from the iconographic record after the Old Kingdom, Egyptians probably continued to use it for larger vessels (see Rogers 1996, 99–104). Ship 17 had a kind of proto-keel that protruded inside the hull (see Belov 2015a) but as it was composed of short segments, it could hardly increase the longitudinal stiffness of the hull to any great degree.
55 J.-P. Olaberria 2015, personal communication. It seems that this function of the through-beams has not yet become the subject of a detailed study.
56 Cooper (2012, 61) cites the late-nineteenth-century data according to which 'the Rashid mouth had a maximum draught of 2.1 m, and the Dumyāt mouth just 1.8 m, compared with 6 m just upstream'.
57 Gaubert and Henein 2015.
58 Collet and Pomey 2015.
59 The dimensions of *lokkafa* of the lake Borollos described by Collet and Pomey has an overall length of 14.5 m, a beam of 5.5 m and a shallow draft of 0.6 m. The boat carries a lateen sail with a windage of 130 m². The average depth of the lake is 1–1.5 m and rarely reaches 2 m.

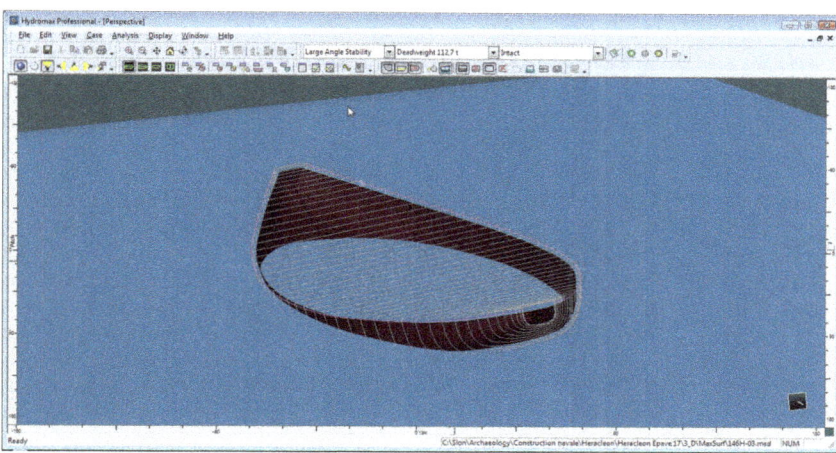

Figure 6.2 Starboard heel of 8 degrees of the hull of ship 17 from Thonis-Heracleion in *Formsys HydroMax*. Loadcase of 113 tons, freeboard of 0.64 m (A. Belov). CC BY 4.0.

The available documentation on a traditional ship from Lake Mareotis (*mariotike*)[60] does not contain any information about their construction, but these ships were probably shallow-draft as well.

6.2.7. Stability

The estimated deadweight of Ship 17 of about 113 tonnes would have resulted in a free-board of 64 cm.[61] The reconstruction of the hull suggests that the ship was very stable with the righting lever (GZ) being maximal at 57 degrees, but in reality the absence of a deck would not permit the heel to exceed 8 degrees (Fig. 6.2.). Apparently, this insufficient heel would only permit navigation on the Nile and in the estuary under good weather conditions.[62] While modestly laden, the ship could

---

60 *BGU* 18.1 2740 dated to 87–86 BC. See Arnaud 2015a, 111.
61 In absence of a deck, this corresponds to the distance from the water to the upper edge of the hull.
62 Difficulties of Nilotic navigation in medieval times are considered in works of Cooper 2008, 2011, 2012, 2012a, 2014 and are corroborated by Ancient Egyptian sources (see Somaglino 2015). Arnaud (2015b, 104, note 14) cites Roman and Late Roman Nilotic contracts with clauses forbidding to navigate at night and in bad weather. Diodorus Siculus (*Bibliotheca Historica* 1.31.2–5) describes the danger of approaching the low coast of Egypt. One should not underestimate complicated navigation in the so-called *boghâz* of which there exist following description left by

probably sail along the coast, although, bearing in mind the strong currents and constant waves at the Nile's estuary,[63] we can define the *barides* mentioned by Herodotus, in connection with the trans-shipment from the Eastern Delta, as decked vessels.[64]

The navigation on the Nile in Antiquity was highly seasonal,[65] and therefore some of the voyages that would not be possible during the flood were possible during the period of low water, and *vice versa*.

6.2.8. Reconstruction of Steering and Propulsion Systems

The evidence for an axial rudder supports the conclusion that the ship's function was of a river or fluvio-maritime type. The first representations of the axial rudder in Egypt have been dated to the end of the Fifth (2494–2345 BC)[66] or to the Sixth Dynasty (2345–2181 BC). This type of rudder was invariably characteristic of Nilotic ships. The boats of type II depicted on the rocks of Rod el-Air, which show parallels with the remains of the seagoing ships from Ayn Sukhna, are seemingly equipped with an axial rudder.[67] However, Pomey notes that these boats were probably adapted to the sea while belonging to the Nilotic boat-building tradition, and that the navigation to the Sinai Peninsula would not have taken more than one day. Generally speaking, an axial rudder did not seem to be a good choice for a sea-going vessel.[68]

---

one of the participants of Napoleon's expedition (1798–1801): 'In Egypt the narrow and perilous straits between the branches of the Nile and the sea are called *boghâz*. These straits are closed by the sands that accumulate due to the confrontation of the high seas with the current of the river. These sandbanks vary depending on the seasons and the winds, so that those bars that are ordinarily found in the mouths of the Nile often change their position, and require the mariners to seek the services of a pilot, who could indicate to them a passage or a channel in the mouth of the river; but this continual surveillance of a pilot is not always sufficient to prevent accidents.' Le Père 1822, 236.

63   See Cooper 2012, 61–62.
64   One of the ships represented on the Nile mosaic from Palestrina in Italy (ancient city of Praeneste) dated to c. 125 BC was recently identified as an example of the *baris* type (Pomey 2015, 164–66). This ship had a large cabin aft of the mast and was probably decked.
65   Among recent publications: Arnaud 2015a, 106–08; 2015b, 8–10; Cooper 2012, 61–64; 2012a, 26; 2014, chapter 7; Somaglino 2015, 127–37.
66   Jones 1995, 39–40.
67   Pomey 2012b, 13; 2012c, 291.
68   A spare rudder was systematically included in the list of Ptolemaic affreightment contracts. Arnaud 2012, 95–96. Fabre (2015, 184, note 47) provides interesting parallels with the axial rudders of the junks.

Figure 6.3 Mortise in the central segment K6 of the proto-keel of ship 17 viewed from above (Photo: C. Gerigk © Franck Goddio/Hilti Foundation).

The masts of Egyptians ships of the period under consideration were situated at the middle of the hull, a conclusion supported by the discovery of a mast-step notch 46 cm long, 13 cm wide and 5 cm deep within Ship 17 of Thonis-Heracleion (Fig. 6.3). Two large mortises in the central strake of the boat Mataria seem to correspond to the middle of the hull and to be related to the position of the mast.[69] The construction of the boat of the Upper Nile *nuggar* may serve as an ethnographic parallel.[70] It has been estimated that the relation between the height of the mast and the length of the hull in the majority of the Egyptian boats must have been close to 2:3.[71] If we accept this ratio, the height of the mast of Ship 17 of Thonis-Heracleion can be estimated at 17–18 m. Obviously, it would have been impossible to obtain a mast of this length from acacia wood, which, according to Herodotus, served as the raw material for its fabrication.[72] Thus two hypotheses may be put forward: either the mast of the *baris* was considerably shorter than if obtained according to the above-mentioned ratio, or it was made from a different wood species. Taking into consideration the precision of Herodotus'

---

69 Haldane 1996, 242.
70 Clarke 1920, 49: 'The stout beam or tree stem was to steady the short mast, which had a socket in the keel and a strap or other form of stay to secure it to the beam.'
71 Goyon 1971, 22.
72 Cf Boreux 1925, 349.

descriptions so far, the initial hypothesis seems more convincing.[73] In case the ship was unable to overcome the current, a mast must have been useful for attaching the tow line.[74]

Figure 6.4 Outer surface of ship 17's keel segment K6 (Photo: Author © Franck Goddio/Hilti Foundation).

6.2.9. The Particular Case of Ship 17 from Thonis-Heracleion

The following arguments are based on evidence from Ship 17 only, and thus they cannot be conclusive. As we shall see, Ship 17 most likely never crossed the sandbar separating the estuary from the sea, but the information at our disposal is still too fragmentary to expand this conclusion on other *barides* from Heracleion.

---

73 Although the meteorology of the Delta was complicated and sometimes resulted in calms (Cooper 2012a, 26), the wind was often favourable for vessels going upriver. According to Arnaud (2015b, 107) the period from December to February was a good time for large ships to sail upstream. The words of Herodotus that a *baris* could not sail upstream unless with a *fresh breeze* may be a slight hint concerning the height of the mast. A short mast would permit hoisting a sail of modest size only and this could explain the difficulty of the ship sailing upstream while being heavily loaded and beating against the strong current.

74 Cf Casson 1965, 36–37, pl. 3, 5.

6.2.10. Traces of Shipworms?

No traces of shipworms were found on the outer surface of the keel, or on the planking of Ship 17 (Fig. 6.4), as in the case of the timbers of seagoing vessels from Mersa Gawasis and Ayn Sukhna.[75] The shipworm *Teredo navalis* can thrive in brackish waters with a salinity as low as 9‰.[76] It seems that Ship 17 was scuttled at the end of its economic life. Even if we forget about constructional limitations and suppose that she made regular sea voyages, this would inevitably result in at least partial infestation by shipworm.

However, it is important to remember that the modern Nile with its regulated run-off radically differs from the ancient river. Thus, even after the construction of the Low Aswan Dam (1902–1933) the river carried approximately 84 km^3 of water in an average year.[77] By 1969, the river had become almost completely auto-controlled and the volume of fresh water reaching the coastal zone dropped to 2.5–4 km^3 per year, in other words by twenty to thirty times less.[78] During the peak of the flood the Nile's run-off was at least eight times higher than during the low water period.[79] Only two of the Nile's branches remain today out of seven that existed during Herodotus' time. Without speculating about intricate paleoclimate models, it is sufficient to cite several modern salinity charts of the region. One can see that even today, the Nile considerably decreases the salinity of the surface layer in the vicinity of the Delta and this phenomenon must have been incomparably more pronounced in Antiquity (Fig. 6.5). It is difficult to calculate whether the salinity dropped to less than 9‰ and the scope of the resulting area of brackish water. This zone probably existed only when the Nile was flooding, and its area would have depended on the circulation pattern of the coastal waters. In conclusion, if ever Ship 17 crossed the sandbar separating the estuary from the coastal waters, she could have stayed there only for a limited period of time before being infested by shipworm.

---

75 I would like to express my gratitude to Prof. David Blackman, who posed this question during the conference 'Heracleion in context' in Oxford in March 2013.
76 *Teredo navalis* can temporarily tolerate salinity of 5 ‰. See Miller 1926, 17.
77 Calculated for the period 1900 to 1959 (White, 1988).
78 Halim and Morcos 1995.
79 See charts in Hurst 1927, 447, Figs. 1–3.

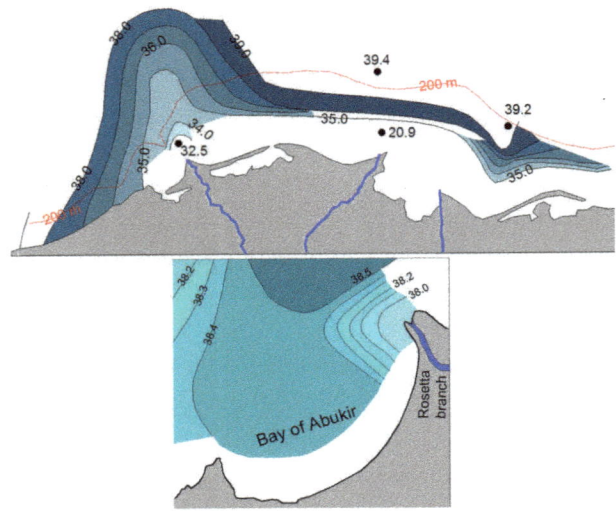

Figure 6.5 Salinity (‰) on the surface of the Mediterranean coast of Egypt in October 1982 and on the surface of the Bay of Abukir in March 1970 (Charts: Author, after Halim & Morcos 1995 [above] and El-Sharkawy & Sharaf el Din 1974 [below]).

6.2.11. Keel's Erosion?

The outer surface of the keel and of the bottom planking of Ship 17 was not eroded at all, and thus there is a slight possibility that the boat was used in another environment than the Delta with its soft muddy banks.

6.3. Conclusions

In the Late Period (664–332 BC), the city of Thonis-Heracleion, 'The Gates of the Sea',[80] controlled access to Egypt, as well as supervising Greek ships in transit to Naucratis[81] and Memphis;[82] it therefore functioned as a customs station and *emporion*.[83] The port commanded

---

80 For the association of this epithet — 'Les Portes de la Mer' — with Thonis see Yoyotte 1994, 683.
81 Yoyotte 1958, 427; 2001, 27.
82 Aramaic papyrus from Saqqara n. 26. Yoyotte 1994, 683; Briant and Descat 1998, 93–95.
83 Fabre and Goddio 2013, 70. Thonis-Heracleion is the most probable site of taxation for the ships mentioned on the Papyrus palimpsest from Elephantine (Ahiqar scroll,

a strategic location at the mouth of one of the most important branches of the Ancient Nile, and perfectly fitted the category of a 'transit point' between 'transport zones' suggested by Westerdahl.[84] The conditions of navigation radically changed at 'transit points', which are usually associated with market places, and this involved 'the reloading of cargo and the change of means of transport at a well-defined site [...] for an accompanying water or land transport in the new zone.'[85] Flat-bottomed vessels from Ostia known as *naves caudicaria*, which were used for the trans-shipment of goods from the sea ports along the river Tiber, serve as a parallel (Fig. 6.6).[86]

Figure 6.6 Mosaic from the Square of Corporations in Ostia representing a scene of transshipment of goods from a sea-going vessel (right) to a riverine navis caudicaria (left) (Photo: I. Sailko, CC BY-SA 3.0, https://commons.wikimedia.org/w/index.php?curid=46365389).

Höckmann argued that there had been regular trans-shipments of cargo from seagoing vessels to river-faring *barides* at Thonis-Heracleion.[87] Villing expressed the opposite point of view; he relied on recent

---

TAD C 3.7, 473–402 BC). See Briant and Descat 1998, 92. The papyrus contains customs registers belonging to thirty-six Ionian and Phoenician ships that came to Egypt in the period from March to December. See Porten and Yardeni 1993.
84 Westerdahl's theory (1992, 6–7) and its application to the mouths of the Nile is discussed by Cooper (2012, 70–71).
85 Westerdahl 1998.
86 Boetto 2001, 2003, 2006, 2008.
87 Höckmann 2008–2009, 78–80, 82–83.

research,[88] which suggested that the Canopic branch in Naucratis 'was wide and deep enough to accommodate Mediterranean seagoing ships all year round; trans-shipment at Thonis-Heracleion, as had sometimes been suspected, was thus not necessary'.[89] It is difficult to agree with this conclusion. True enough, written sources[90] are corroborated by several maritime finds from Naucratis.[91] Thomas estimates the Canopic branch to be 5 m deep and circa 200 m wide near urban areas and concludes that '[the] Canopic branch of the Nile was deep enough and navigable, likely all year round, for sea-going ships such as the Kyrenia'.[92] However, we should note that the Kyrenia was a fourth-century-BC Greek merchant ship of a very modest size. This 25-tonner's reconstructed length was about 13.86 m and she had a loaded draft of only 1.47 m.[93] Apparently this ship would not have encountered serious problems in reaching Naucratis, even during the low-water season.

Seagoing ships of considerable tonnage regularly came upriver to Naucratis, Memphis and even Thebes; thus, trans-shipment was not an obligation. Indeed, the loaded draft of the larger ships did not exceed the depth of the Canopic branch.[94] Nantet shows that, in the Hellenistic period, large ships with tonnage exceeding 10,000 *artabs* (200–250 tonnes) and reaching as much as 18,000 *artabs* (about 400 tonnes), were numerous on the Nile.[95] Arnaud notices that these ships, called *kerkouroi*, surely came from the sea and that they were more numerous in winter, from December to February, than during the high-water season.[96] However, it is essential to remember that these ships operated from Alexandria and so did not need to enter the mouths of the Nile, a task that was sometimes difficult to do in earlier periods.

The sources also tell us that military fleets were able to come upriver. One of the insurgencies during the first Persian domination

---

88 Pennigton and Thomas, in preparation; Thomas 2015.
89 Villing 2015, 231.
90 See note 79.
91 Thomas 2015, 253, Fig. 13.5.
92 Ibid., 252.
93 Tonnage: Steffy 1985, 100. Draft: Katzev 1981, 318.
94 Nantet lists thirteen shipwrecks, with their deadweight tonnages and their loaded drafts. All of them could have sailed on the Canopic branch. Even the Madrague de Giens shipwreck, with a deadweight tonnage of 402.5 tonnes and a loaded draft of 3.75 m, could have sailed there. Nantet 2016, 226, Table 47.
95 Nantet 2016, 575–76. See also chapter 1 in this book.
96 Arnaud 2015b, 106–09, 112.

(Twenty-seventh Dynasty, 525–404 BC) was supported by an Athenian fleet of 200 triremes that sailed up the Nile and seized the larger part of Memphis (Thucydides 1.104). This is not surprising, since these warships were relatively light and shallow-drafted. The replica of the Athenian trireme 'Olympias' had a draft of only 1.1 m.[97]

Returning to the trade vessels of the Late Period, there is important evidence from the Persian era — the Ahiqar scroll[98] — that contains a list of foreign ships that passed through an unnamed port on the Delta, most probably Thonis-Heracleion.[99] There are two major groups of ships on the list: the 'small' ships had a tonnage of about 40 tonnes and the 'large' ones about 60 tonnes.[100] Here again one is dealing with relatively small merchant ships that could easily come upriver on their own.[101]

As far as the *barides* of Thonis-Heracleion were concerned, their construction strongly indicates a river origin of this type, which probably had a fluvio-maritime designation.[102] If the assumption of the hull's form from Ship 17 is correct, they were well-adapted to navigation within the estuary but were not particularly seaworthy.

It has been suggested that these ships could have belonged to the temple fleet.[103] Several possibilities can be proposed for their use: either they were involved in the trans-shipment of goods from the larger seagoing vessels that could not enter Thonis-Heracleion because of their considerable draft, or they transported goods from Heracleion up the river, or, finally, both. The absence of a deck would have been a definite advantage for rapid trans-shipment.

In my opinion, the question of trans-shipment at Heracleion is not as unambiguous as it is sometimes assumed to be.[104] There are too many parameters involved for us to be certain: the tonnage of the ship, the nature of her cargo, seasonality and meteorological conditions, etc. Some of the seagoing ships continued their journey upriver on their

---

97 Morrison 1996, 345; Morrison et al. 2000, 156.
98 See note 80.
99 Höckmann, 110.
100 Briant and Descat 1998, 68. See also Nantet 2016, 575.
101 According to Wilson 2011 (39, note 27) 'ships of less than 75 tons were common throughout the Roman period as they were before and afterwards' [the long ton of 1016 kg is used — AB].
102 Arnaud 2015b, 116.
103 Robinson 2015, 222, 291, note 51.
104 Cf Höckmann 2008–2009, 83; Villing 2015, 231.

own while others needed trans-shipment to the river-faring craft. It was only the question of tonnage and draft: as evidenced by Herodotus (2.179, see above) it was sometimes difficult to navigate around the Delta due to the contrary winds. On the other hand, the Delta often had calm waters, and that could also have necessitated trans-shipment.[105]

Bibliography

Literary Sources

Achilles Tatius, *Leucippe and Clitophon.*

Diodorus Siculus, *Bibliotheca Historica.*

Heliodorus, *Aethiopica.*

Herodotus, *Historiae.*

Thucydides, *History of the Peloponnesian War.*

Secondary Sources

Abd el-Maguid, M. 2015. 'An Elongated Composite Stone Anchor from Matariya.' In *Alexandria Under the Mediterranean. Archaeological Studies in Memory of Honor Frost*, edited by G. Soukiassian, 125–36. Etudes Alexandrines 36. Alexandria: Centre d'Études Alexandrines.

Arnaud, P. 2012. 'La mer, vecteur des mobilités grecques.' In *Mobilités grecques*, edited by L. Capdetrey, 89–135. Scripta Antiqua 46. Bordeaux: Ausonius.

Arnaud, P. 2015a. 'Navires et navigation commerciale sur la mer et sur le "Grand fleuve" à l'époque des Ptolémées.' In *Entre Nil et mers. La navigation en Égypte ancienne*, edited by B. Argémi and P. Tallet, 105–122. *Nehet* 3

Arnaud, P. 2015b. 'La batellerie de fret nilotique d'après la documentation papyrologique (300 av. J.-C.–400 apr. J.-C.).' In *La batellerie Egyptienne. Archéologie, histoire, ethnographie*, edited by P. Pomey, 99–150. Études Alexandrines 34. Paris: Centre d'Études Alexandrines.

---

105 Cooper 2012a, 26. 'The warming land often generated land breezes that cancelled out the prevailing northwesterlies blowing in off the Mediterranean Sea, resulting in frequent calms that left Nile boats facing a strong current with no supporting wind. Even when the sea breezes broke through, Delta channels were not always oriented in a way that enabled navigators to take advantage of them. The combined result of all these factors was hard labour for Nile navigators.'

Basch, L. 1985. 'Anchors in Egypt.' *Mariners Mirroir* 71: 453–67.

Basch, L. 1987. *Le musée imaginaire de la marine antique*. Athens: Institut Hellénique pour la Préservation de la Tradition Nautique.

Basch, L. 1994. 'Some Remarks on the Use of Stone Anchors and Pierced Stones in Egypt.' *International Journal of Nautical Archaeology* 23(3): 219–27.

Belov, A. 2014. 'A New Type of Construction Evidenced by Ship 17 of Heracleion-Thonis.' *International Journal of Nautical Archaeology* 43(2): 314–29. https://doi.org/10.1111/1095-9270.12060

Belov, A. 2015a. 'Did Ancient Egyptian Ships have Keels? The Evidence of Thonis-Heracleion Ship 17.' *International Journal of Nautical Archaeology* 44(1): 74–80. https://doi.org/10.1111/1095-9270.12078

Belov, A. 2015b. 'Archaeological Evidence for the Egyptian *baris* (Herodotus, II.96).' In *Thonis-Heracleion in Context: The Maritime Economy of the Egyptian Late Period*, edited by D. Robinson and F. Goddio, 195–210. Oxford Center for Maritime Archaeology Monographs 8. Oxford: Oxford Center for Maritime Archaeology.

Belov, A. 2019. 'Ship 17: A Late Period Egyptian Ship from Thonis-Heracleion.' In *Oxford Centre for Maritime Archaeology Monograph 10*. Oxford: Oxford Centre for Maritime Archaeology.

Bird, E. C. F. 1994. 'Physical Setting and Geomorphology of Coastal Lagoons.' In *Coastal Lagoon Processes*, edited by B. Kjerfve, 9–39. Elsevier Oceanography Series 60. Amsterdam, London, New York and Tokyo: Elsevier.

Boreux, C. 1925. *Etudes de nautique égyptienne: l'art de la navigation en Egypte jusqu'à la fin de l'ancien Empire*. Cairo: Institut français d'archéologie orientale.

Casson, L. 1965. 'Harbour and River Boats of Ancient Rome.' *Journal of Roman Studies* 55: 31–9.

Casson, L. 1971. *Ships and Seamanship in Ancient World*. Princeton: Princeton University Press.

Cataudella, S., D. Crosetti, and F. Massa, eds. 2015. *Mediterranean Coastal Lagoons. Sustainable Management and Interactions among Aquaculture, Capture Fisheries and the Environment, Studies and Reviews* 95. Rome: Food and Agriculture Organization of the United Nations. General Fisheries Commission for the Mediterranean.

Clarke, S. 1920. 'Nile Boats and Other Matters.' *Ancient Egypt*, 2–50.

Collet, R., and P. Pomey. 2015. 'Les voiles de Borollos.' In *La batellerie Egyptienne. Archéologie, histoire, ethnographie*, edited by P. Pomey, 299–314. Etudes Alexandrines 34. Paris: Centre d'Études Alexandrines.

Cooper, J. P. 2008. 'The Medieval Nile: Route, Navigation and Landscape in Islamic Egypt.' PhD diss., University of Southampton.

Cooper, J. P. 2011. 'No Easy Option: Nile Versus Red Sea in Ancient and Medieval North-South Navigation.' In *Maritime Technology in the Ancient Economy: Ship-Design and Navigation*, edited by W. V. Harris and K. Iara, 189–210. *Journal of Roman Archaeology* Suppl 84.

Cooper, J. P. 2012. "Fear God; Fear the Bogaze': The Nile Mouths and the Navigational Landscape of the Medieval Nile Delta, Egypt.' *Al-Masaq: Islam and the Medieval Mediterranean* 24(1): 53–73. https://doi.org/10.1080/09503110.2012.655584

Cooper, J. P. 2012a. 'Nile Navigation: Towing All Day, Punting for Hours.' *Egyptian Archaeology* 41: 25–27.

Cooper, J. P. 2014. *The Medieval Nile: Route, Navigation, and Landscape in Islamic Egypt*. Cairo: The American University in Cairo Press. https://doi.org/10.5743/cairo/9789774166143.001.0001

Darnell, J. C. 1992. 'The Kbn.wt vessels of the Late Period.' In *Life in a Multicultural Society: Egypt from Cambyses to Constantine and Beyond*, edited by J. H. Johnson, 67–89. Studies in Ancient Oriental Civilization 51. Chicago: Oriental Institute of the University of Chicago.

Doyle, N. 1998. 'Iconography and Interpretation of Ancient Egyptian Watercraft.' M.A. diss., Texas A&M University.

El-Sharkawy, S. H., and S. H. Sharaf el Din. 1974. 'Hydrographic Structure and Circulation Pattern of Abu Kir Bay Near Alexandria, Egypt.' *Bulletin of the Institute of Oceanography and Fisheries, Egypt* 4: 461–71.

El-Wakeel, S. K., and S. D. Wahby. 1970. 'Bottom Sediments of Lake Manzalah, Egypt: Notes.' *Journal of Sedimentary Petrology* 40(1): 480–96.

Fabre, D. 2008. 'Heracleion-Thonis: Customs Station and Emporion.' In *Egypt's Sunken Treasures. Catalogue of the Exhibition in Madrid*, edited by F. Goddio and M. Clauss, 219–25. Munich, Berlin, London and New York: Prestel.

Fabre, D. 2015. 'The Ships of Thonis-Heracleion in Context.' In *Thonis-Heracleion in Context: The Maritime Economy of the Egyptian Late Period*, edited by D. Robinson and F. Goddio, 175–94. Oxford: Oxford Center for Maritime Archaeology.

Fabre, D. and Belov, A. 2012. 'The Shipwrecks of Heracleion-Thonis: An Overview.' In *Achievements and Problems of Modern Egyptology. Proceedings of the International Conference*. 29 September–4 October 2009, Moscow, edited by G. A. Belova, 107–18. Moscow: Russian Academy of Sciences.

Fabre, D. and F. Goddio. 2013. 'Thonis-Heracleion, Emporion of Egypt, Recent Discoveries and Research Perspectives: The Shipwrecks.' *Journal of Ancient Egyptian Interconnections* 5(1): 68–75.

Frost, H. 1970. 'Bronze Age Stone Anchors from the Eastern Mediterranean.' *The Mariner's Mirror* 56: 377–94.

Frost, H. 1985. 'The Kition Anchors.' In *Kition V: The Pre-Phoenician Levels*, edited by V. Karageorghis and M. Demas, 295–306. Nicosia: Department of Antiquities.

Frost, H. 1995. 'Where Did Bronze Age Ships Keep their Stone Anchors?' In *TROPIS III. 3nd International Symposium on Ship Construction in Antiquity. Proceedings*, edited by H. Tzalas, 225–29. Athens: Hellenic Institute for the Preservation of Nautical Tradition.

Gaubert, C., and N. H. Henein. 2015. 'Le bateau du lac Manzala.' In *La batellerie Egyptienne. Archéologie, histoire, ethnographie*, edited by P. Pomey, 285–98. Etudes Alexandrines 34. Alexandrie, Centre d'Études Alexandrines.

Goedicke, H. 1975. *The Report of Wenamun*. Baltimore and London: Johns Hopkins University Press.

Goddio, F. 2007. *The Topography and the Excavation of Heracleion-Thonis and East Canopus (1996–2006). Vol. 1, Underwater Archaeology in the Canopic Region in Egypt*. Oxford: Oxford University Press for the British Academy.

Goddio, F. 2011. 'Heracleion-Thonis and Alexandria, Two Ancient Egyptian Emporia.' In *Maritime Archaeology and Ancient Trade in the Mediterranean*, edited by D. Robinson, 121–38. Oxford Center for Maritime Archaeology Monograph 6. Oxford: Oxford Center for Maritime Archaeology.

Goddio, F., D. Robinson, and D. Fabre. 2015. 'The Life-cycle of the Harbour of Thonis-Heracleion: The Interaction of the Environment, Politics and Trading Networks on the Maritime Space of Egypt's Northwestern Delta.' In *Harbours and Maritime Networks as Complex Adaptive Systems*, edited by J. Preiser-Kapeller and F. Daim, 25–39. Mainz: Verlag des Römisch-Germanischen Zentralmuseums.

Goddio, F. 2015. 'The Sacred Topography of Thonis-Heracleion.' In *Thonis-Heracleion in Context: The Maritime Economy of the Egyptian Late Period*, edited by D. Robinson and F. Goddio, 15–54. Oxford: Oxford Center for Maritime Archaeology.

Goyon, G. 1971. 'Les navires de transport de la chaussée monumentale d'Ounas.' *Bulletin de l'institut français d'archéologie orientale* 69: 11–41.

Haldane, C. W. 1993. 'Ancient Egyptian Hull Construction.' PhD diss., Texas A&M.

Haldane, C. W. 1996. 'Ancient Egyptian Hull Construction.' In *TROPIS IV. Proceedings of the 4th International Symposium on Ship Construction in Antiquity*, edited by H. Tzalas. Athens.

Halim, Y., and S. A. Morcos. 1995. 'The Impact of the Nile and the Suez Canal on the Living Marine Resources of the Egyptian Mediterranean Waters (1958–1986).' In *Effects of Riverine Inputs on Coastal Ecosystems and Fisheries Resources*. FAO Fisheries Technical Paper, 19–50. Rome: Food and Agriculture Organization of the United Nations.

Hirth, K. G. 1978. 'Interregional Trade and the Formation of Prehistoric Gateway Communities.' *American Antiquity* 43: 35–45.

Höckmann, O. 2008–2009. 'Griechischer Seeverkehr mit dem archaischen Naukratis in Agypten.' *Talanta* 40–41: 73–135.

Hornell, J. 1943. 'The Sailing Ship in Ancient Egypt.' *Antiquity* 17: 27–41.

Hurst, H. E. 1927. 'Progress in the Study of the Hydrology of the Nile in the Last Twenty Years.' *The Geographical Journal* 70(5): 440–58.

Janssen, J. 1975. *Commodity Prices from the Ramesside Period*. Leiden: E. J. Brill.

Jones, D. 1988. *A Glossary of Ancient Egyptian Nautical Titles and Terms*. London & New York: Kegan Paul International.

Jones, D. 1995. *Egyptian Bookshelf: Boats*. London: British Museum.

Katzev, M. L. 1981. 'The Reconstruction of the Kyrenia Ship, 1972–1975.' *National Geographic Society Research Reports* 13: 315–28.

Le Père, J. M. 1822. 'Mémoire sur le canal.' In *Description de l'Égypte: ou, Recueil des observations et des recherches qui ont été faites en Egypte pendant l'expédition de l'armée française, publié par les ordres de Sa Majesté l'empereur Napoléon le Grand*, edited by J. Jollios, 236–39. Paris: Imprimerie Impériale.

Macaulay, G. C. 1890. *The History of Herodotus*. London: Macmillan.

Miller, R. C. 1926. 'Ecological Relations of Marine Wood-Boring Organisms in San Francisco Bay.' *Ecology* 7(3): 247–54.

Morrison, J. S. 1996. *Greek and Roman Oared Warships 399–30 BC*. Oxford: Oxbow Books.

Morrison, J. S., J. F. Coates, and N. B. Rankov. 2000. *The Athenian Trireme. The History and Reconstruction of an Ancient Greek Warship*. 2nd ed. Cambridge: Cambridge University Press.

Nantet, E. 2016. *Phortia. Le tonnage des navires de commerce en Méditerranée du VIIIe siècle av. l'ère chrétienne au VIIe siècle de l'ère chrétienne*. Rennes: Presses Universitaires de Rennes.

Nibbi, A. 1991. 'Five Stone Anchors from Alexandria.' *International Journal of Nautical Archaeology and Underwater Exploration* 20: 185–94.

Pennington, B., and I. R. Thomas. In preparation. 'Palaeo-Landscape Reconstruction at Naukratis and the Canopic Branch of the Nile.'

Pomey, P. 2012a. 'Ship Remains at Ayn Soukhna.' In *The Red Sea in Pharaonic Times: Recent Discoveries Along the Red Sea Coast. Proceedings of the Colloquium Held in Cairo/Ayn Sukhna 11–12 January 2009*, edited by P. Tallet, 235–44. Cairo: Institut Français d'Archéologie Orientale.

Pomey, P. 2012b. 'The Pharaonic Ship Remains of Ayn Sukhna.' In *Between Continents: Proceedings of the Twelfth Symposium on Boat and Ship Archaeology*

(ISBSA 12 — Istanbul 2009), edited by N. Günsenin: 7–15. Istanbul: Ege Yayınları.

Pomey, P. 2012c. 'Les graffiti navals de la zone minière du Sud-Sinaï (Rod al-Air, Gebel al-Hazbar, Ouadi Shella/Ouadi Boudra).' In *La zone minière pharaonique du Sud-Sinaï- I. Catalogue complémentaire des inscriptions du Sinaï*, edited by P. Tallet. MIFAO 130. Cairo: Institut Français d'Archéologie Orientale.

Pomey, P. 2015. 'La batellerie nilotique gréco-romaine d'après la mosaïque de Palestrina.' *La batellerie Egyptienne. Archéologie, histoire, ethnographie*, edited by P. Pomey 151–73. Alexandria: Centre d'Études Alexandrines.

Porten, B., and A. Yardeni. 1993. *Textbook of Aramaic Documents from Ancient Egypt. 3. Literature, Accounts, Lists*. Jerusalem: Hebrew University.

Robinson, D., and F. Goddio, eds. 2015. *Thonis-Heracleion in Context: The Maritime Economy of the Egyptian Late Period*. Oxford Centre for Maritime Archaeology 8. Oxford: Oxford Centre for Maritime Archaeology.

Robinson, D. 2015. 'Ship 43 and the Formation of the Ship Graveyard in the Central Basin at Thonis-Heracleion.' In *Thonis-Heracleion in Context: The Maritime Economy of the Egyptian Late Period* edited by D. Robinson and F. Goddio, 211–27. Oxford: Oxford Center for Maritime Archaeology.

Robinson, D. 2018. 'The Depositional Contexts of the Ships from Thonis-Heracleion, Egypt.' *International Journal of Nautical Archaeology* 47(2): 325–36, https://doi.org/10.1111/1095-9270.12321

Rogers, E. M. 1996. 'An Analysis of Tomb Reliefs Depicting Boat Construction from the Old Kingdom Period in Egypt.' PhD diss., Texas A&M University.

Somaglino, C. 2015. 'La navigation sur le Nil. Quelques réflexions autour de l'ouvrage de J. P. Cooper, The Medieval Nile. Route, Navigation, and Landscape in Islamic Egypt, Le Caire — New York, 2014.' In *Entre Nil et mers. La navigation en Égypte ancienne*, edited by B. Argémi and P. Tallet, 123–61. *Nehet* 3. Paris: Centre de recherches égyptologiques de la Sorbonne.

Steffy, J. R. 1985. 'The Kyrenia Ship: An Interim Report on its Hull Construction.' *American Journal of Archaeology* 89(1): 71–101.

Tallet, P. 2013. 'The Wadi el-Jarf site: A Harbor of Khufu on the Red Sea.' *Journal of Ancient Egyptian Interconnections* 5(1): 76–84.

Tallet, P. 2015. 'Les 'ports intermittents' de la mer Rouge à l'époque pharaonique: caractéristiques et chronologie.' In *Entre Nil et mers. La navigation en Égypte ancienne*, edited by B. Argémi and P. Tallet, 31–72. *Nehet* 3. Paris: Centre de recherches égyptologiques de la Sorbonne.

Thomas, I. R. 2015. 'Naukratis, "Mistress of ships", In Context.' In *Thonis-Heracleion in Context: The Maritime Economy of the Egyptian Late Period*, edited by D. Robinson and F. Goddio, 247–65. Oxford Centre for Maritime Archaeology, Monograph 8. Oxford: Oxford Centre for Maritime Archaeology.

Thompson, D. 1983. 'Nile Grain Transports under the Ptolemies.' In *Trade in the Ancient Economy*, edited by P. Garnsey, K. Hopkins, and C. R. Whittaker, 64–75. Berkeley and Los Angeles: University of California Press.

Tzalas, H. 1999. 'Were the Pyramidal Stone-Weights of Zea Used as Anchors?' In *TROPIS V — 5th International Symposium on Ship Construction in Antiquity*, edited by H. Tzalas, 429–54. Athens: Hellenic Institute for the Preservation of Nautical Tradition.

Villing, A. 2015. 'Egyptian-Greek Exchange in the Late Period: The View from Nokradj-Naukratis.' In *Thonis-Heracleion in Context: The Maritime Economy of the Egyptian Late Period*, edited by D. Robinson and F. Goddio. Oxford: Oxford Center for Maritime Archaeology.

Vinson, S. 1994. *Egyptian Boats and Ships*. Buckinghamshire: Shire Publications.

Vinson, S. 1997. 'On *hry.t* 'Bulwark', in P. Anastasi IV, 7/9–8/7.' *Zeitschriff fur Agyptische Sprache* 124(2): 156–62.

Vinson, S. 1998. 'Remarks on Herodotus' Description of Egyptian Boat Construction (II, 96).' *Studien zur Altägyptischen Kultur* 26: 251–60.

Ward, C. 2004. 'Boatbuilding in Ancient Egypt.' In *The Philosophy of Shipbuilding. Conceptual Approaches to the Study of Wooden Ships*, edited by F. Hocker and C. Ward, 12–24. College Station: Texas A&M University Press.

Ward., C. 2007. 'Ship-Timbers: Description and Preliminary Analysis.' In *Harbor of the Pharaohs to the Land of Punt*, edited by K. A. Bard, 135–50. Naples: Università degli Studi di Napoli 'L'Orientale.'

Ward, C., and Zazzaro, C. 2010. 'Evidence for Pharaonic Seagoing Ships at Mersa/Wadi Gawasis, Egypt.' *International Journal of Nautical Archaeology* 39 (1): 27–44. https://doi.org/10.1111/j.1095-9270.2009.00229.x

Wehausen, J. V. 1988. 'The Colossi of Memnon and Egyptian Barges.' *International Journal of Nautical Archaeology and Underwater Exploration* 17(4): 295–310.

Westerdahl, C. 1992. 'The Maritime Cultural Landscape.' *International Journal of Nautical Archaeology and Underwater Exploration* 21 (1): 5–14.

Westerdahl, C. 1998. 'The Maritime Cultural Landscape: On the Concept of the Traditional Zones of Transport Geography.' https://www.abc.se/~m10354/publ/cult-land.htm

Wilson, A. 2011. 'Developments in Mediterranean Shipping and Maritime Trade from the Hellenistic Period to AD 1000.' In *Maritime Archaeology and Ancient Trade in the Mediterranean*, edited by D. Robinson, 33–60. Oxford Center for Maritime Archaeology Monograph 6. Oxford: Oxford Centre for Maritime Archaeology.

White, G. F. 1988. 'The Environmental Effects of the High Dam at Aswan.' *Environment* 30: 5–40.

Yoyotte, J. 1958. 'Notes de toponymie égyptienne.' In *Festschrift zum 80. Geburtstag von Professor Dr. Hermann Junker, Mitteilungen des Deutschen Archäologischen Instituts, Abteilung Kairo* 16: 414–30. Wiesbaden: Otto Harrassowitz.

Yoyotte, J. 1994. 'Les contacts entre Egyptiens et Grecs (VIeI–IIe s. av.J.-C.): Naucratis, ville égyptienne.' *Annuaire du Collège de France. Résumé des cours et travaux* (1992–1993, 1993–1994): 679–92.

Yoyotte, J. 2001. 'Le second affichage du décret de l'an 2 de Nekhtnebef et la découverte de Thônis-Héracleion.' *Egypte, Afrique & Orient* 24: 24–34.

Yoyotte, J. 2013. *Histoire, géographie et religion de l'Egypte ancienne: Opera Selecta*. Leuven: Peeters Publishers.

Zazzaro, C. 2007. 'Ship Blades, Anchors and Pierced Stones.' In *Harbour of the Pharaohs to the Land of Punt*, edited by K. A. Bard and and R. Fattovich, 150–60. Naples: Università di Napoli 'L'Orientale'.

Zazzaro, C. 2011. 'Les ancres de Mersa Gawasis.' *Egypte, Afrique & Orient* 64(2): 13–20.

Zazzaro, C., and M. Abd el-Maguid. 2012. 'Ancient Egyptian Stone Anchors from Mersa Gawasis.' In *The Red Sea in Pharaonic Times: Recent Discoveries along the Red Sea Coast. Proceedings of the Colloquium held in Cairo/Ayn Sukhna 11–12 January 2009*, edited. by P. Tallet, 87–103. BIFAO 155. Cairo: IFAO.

List of Tables and Illustrations

Chapter 2

Fig. 2.1   Itineraries mentioned in the *Stadiasmus* (CAD Anne-Laure Pharisien/CReAAH). 15
Fig. 2.2   Itineraries mentioned in Strabo's *Geography* (CAD Anne-Laure Pharisien/CReAAH). 15
Fig. 2.3   Weather conditions around Cyprus in December (CAD Anne-Laure Pharisien/CReAAH). 19
Fig. 2.4   Weather conditions around Cyprus in June (CAD Anne-Laure Pharisien/CReAAH). 19
Table 2.1 Comparison of the impact of weather conditions on navigation. 21
Fig. 2.5   *Akrai* and *akroteria* in Strabo's *Geography* (CAD Anne-Laure Pharisien/CReAAH). 22
Fig. 2.6   *Akrai* and *akroteria* in the *Stadiasmus* (CAD Anne-Laure Pharisien/CReAAH). 22
Table 2.2 Main *akroteria* and *akrai* mentioned in the *Stadiasmus* and in Strabo's *Geography*. 23

Chapter 3

Fig. 3.1   Kyrenia shipwreck. Plan and amidship cross-section (Steffy 1994). 30
Fig. 3.2   Kyrenia shipwreck. Reconstructed hull lines (Steffy 1994). 31
Fig. 3.3   Marsala shipwreck. Hull plan (Frost 1976). 32
Fig. 3.4   Madrague de Giens shipwreck. General view of hull (Photo A. Chéné, AMU, CNRS, MCC, CCJ). 32
Fig. 3.5   Madrague de Giens shipwreck. Plan of the hull remains (Drawing M. Rival, AMU, CNRS, MCC, CCJ). 33

Fig. 3.6 Madrague de Giens shipwreck. Reconstructed hull lines (Drawing M. Rival, AMU, CNRS, MCC, CCJ). 34

Fig. 3.7 Madrague de Giens shipwreck. Amidship cross-sections (Drawing J.-M. Gassend, M. Rival, AMU, CNRS, MCC, CCJ). 34

Fig. 3.8 Madrague de Giens shipwreck. General axonometric view (Drawing J.-M. Gassend, M. Rival, AMU, CNRS, MCC, CCJ). 35

Fig. 3.9 Madrague de Giens shipwreck. Axial axonometric view (Drawing J.-M. Gassend, M. Rival, AMU, CNRS, MCC, CCJ). 35

Fig. 3.10 Madrague de Giens shipwreck. Axonometric view of the keel, the double planking and the hull sheathing (Drawing J.-M. Gassend, M. Rival, AMU, CNRS, MCC, CCJ). 36

Fig. 3.11 Madrague de Giens shipwreck. Axonometric views of the stem complex and the stern complex (Drawing M. Rival, AMU, CNRS, MCC, CCJ). 37

Fig. 3.12 Madrague de Giens shipwreck. 3D reconstruction of the hull shapes (Drawing Sistre international). 38

Fig. 3.13 Madrague de Giens shipwreck. Detailed section of the keel area. Note the bolt joining the floor-timber to the keel (Drawing J.-M. Gassend, M. Rival, AMU, CNRS, MCC, CCJ). 38

Fig. 3.14 Dramont A shipwreck. Axonometric view of the central part of the hull (Drawing Cl. Santamaria). 39

Fig. 3.15 Jules-Verne 9 shipwreck. General view of the hull remains (Photo M. Derain, AMU, CNRS, MCC, CCJ). 40

Fig. 3.16 Jules-Verne 9 shipwreck. Cross-section of the hull remains (Drawing M. Rival, AMU, CNRS, MCC, CCJ). 40

Fig. 3.17 Jules-Verne 9 shipwreck. Axonometric view of the sewing and lashing of the hull assembly system (Drawing M. Rival, AMU, CNRS, MCC, CCJ). 41

Fig. 3.18 Jules-Verne 7 shipwreck. General view of the hull remains (Photo M. Derain, AMU, CNRS, MCC, CCJ). 41

Fig. 3.19 Jules-Verne 7 shipwreck. Amidship cross-section of the hull remains (Drawing M. Rival, AMU, CNRS, MCC, CCJ). 42

Fig. 3.20 Theoretical schema of the mortise-and-tenon joint (Drawing M. Rival, AMU, CNRS, MCC, CCJ). 42

Fig. 3.21 Jules-Verne 7 shipwreck. Schema of the mortise-and-tenon joint network (Drawing M. Rival, AMU, CNRS, MCC, CCJ). 42

Fig. 3.22 Jules-Verne 7 shipwreck. General axonometric view of the hull structure (toward the bow) (Drawing M. Rival, AMU, CNRS, MCC, CCJ). 43

Fig. 3.23 Ma'agan Mikhael shipwreck. Plan and longitudinal section of the hull remains (Kahanov, Linder 2004). 44

List of Tables and Illustrations 121

Fig. 3.24 Ma'agan Mikhael shipwreck. Main cross-section of the hull remains (From Kahanov, Linder 2004). 45
Fig. 3.25 Ma'agan Mikhael shipwreck. Top view of the bow with the sewn bow knee (Kahanov, Linder 2004). 45
Fig. 3.26 Trireme replica *Olympias*. General plans (J.F. Coates). 46
Fig. 3.27 a- Baie de Briande shipwreck; b- Chrétienne A shipwreck. (Drawing M. Rival, AMU, CNRS, MCC, CCJ). 47
Fig. 3.28 Western Roman Imperial type: top- Laurons 2 shipwreck; bottom- *La Bourse* shipwreck (Marseilles) (P. Pomey, AMU, CNRS, MCC, CCJ). 48
Fig. 3.29 Yassiada 2 shipwreck. Cross-sections at frame B7 and B23 (van Doorninck 1976). 48
Fig. 3.30 Yassiada 1 shipwreck. Amidship cross-sections (Steffy 1982). 49
Fig. 3.31 Bozborum shipwreck. Cross-section of the hull (floor-timber 1) (Harpster 2002). 49
Fig. 3.32 Mosaic of the *frigidarium* of the bath of Themetra (Tunisia, 3rd c. AD). Ship of Madrague de Giens type (Photo R. Guéry, AMU, CNRS, MCC, CCJ). 50
Fig. 3.33 Comparative sketch of the Themetra ship and the Madrague de Giens. Note the similarity of the hull profiles (Drawing M. Rival, AMU, CNRS, MCC, CCJ). 50
Fig. 3.34 Mosaic of the *Syllectani* in the *Piazzale delle Corporazioni* (Ostia Antica, late 2nd c. AD) (Photo A. Chéné, AMU, CNRS, MCC, CCJ). 51

Chapter 5

Fig. 5.1 The evolution of the tonnage of the ships in the Hellenistic period from the shipwrecks. Graph by Emmanuel Nantet. CC BY. 76
Table 5.1 The tonnage of the ships in the Hellenistic period from the shipwrecks. 77
Fig. 5.2 The evolution of the tonnage of the boats mentioned in the papyri in the Hellenistic period. Graph by Emmanuel Nantet. CC BY. 78
Table 5.2 Estimated number of shipments required for the supply of Rome. 81
Table 5.3 The gifts of the Western Mediterranean powers to the Romans from the second half of the 3rd century to the first half of the 2nd century BCE (after Garnsey 1996, 241-246). 82

Chapter 6

Fig. 6.1 Simplified topography of the Canopic region (After Goddio 2007, 17, fig. 1.15.)   92

Fig. 6.2 Starboard heel of 8 degrees of the hull of ship 17 from Thonis-Heracleion in *Formsys HydroMax*. Loadcase of 113 tons, freeboard of 0.64 m. CC BY 4.0.   101

Fig. 6.3 Mortise in the central segment K6 of the proto-keel of ship 17 viewed from above (Photo: C. Gerigk © Franck Goddio/Hilti Foundation).   103

Fig. 6.4 Outer surface of ship 17's keel segment K6 (Photo: Author © Franck Goddio/Hilti Foundation).   104

Fig. 6.5 Salinity (‰) on the surface of the Mediterranean coast of Egypt in October 1982 and on the surface of the Bay of Abukir in March 1970 (Charts: Author, after Halim & Morcos 1995 and El-Sharkawy & Sharaf el Din 1974).   106

Fig. 6.6 Mosaic from the Square of Corporations in Ostia representing a scene of transshipment of goods from a sea-going vessel to a riverine navis caudicaria (Photo: I. Sailko, CC BY-SA 3.0, https://commons.wikimedia.org/w/index.php?curid=46365389).   107

Index

Achaeans 13, 57
Actium 59, 67–68
Aegean Sea 2, 6, 13, 78, 85
Ahiqar scroll 109
Albenga 39, 83
Alexander the Great 55, 59
  Alexandrian Empire 28
Alexandria 3, 13, 78, 81, 91, 108
al-Itritus 23
Amathus 18
Ammochostos 18
amphorae 2–3, 29, 84
anchorage 23
Ancient Egyptian 91, 93, 96, 98–100
Ancient period 11, 56, 64, 91, 93, 96–100, 107
Antikythera 77, 83, 85
Antiquity xii, 66, 97, 102, 105
Antirhodos 84
Apelles xv, 55, 59, 65
Aphrodite 13–14, 24
Apollonia 2, 77
archaeology xi, xii, xiii, xiv, xv, 2–3, 5–6, 40, 78, 83, 96, 98
Archaic Greek xiii
Archaic period xi, xii, xiii, 1, 13, 29, 75
Archimedes 28
Aristonothos krater 61
Arnaud, Pascal xiv, 78, 108
Arsinoe 18
Asia Minor 6, 13
Athenaeus 63–64
Athenian xvi, 58, 109
Athens xvi, 58–59, 62, 65–67, 80

Athlit ram 62
Attic works 61
axial frame 29
Ayia Napa 5
Ayn Sukhna 102, 105

Baie de Briande 39, 46
baris xiv, xv, 6, 91, 93–95, 98–100, 103
basin 14, 80
Bay of Abukir 91
Berenice 63
Berenike II 62
Bon-Porté 40, 45
bow 30, 62–63, 68, 96
Bozburun 47
breakwaters 14
Bronze Age 46

Caesar 1 41
Caesarea 84
Cala Sant Vicenç 40
Cap Bénat 84
Cape Greco 24
Cape Pedalion 24
cargo 3–4, 81, 83–84, 86
Cavalière 39
Caveaux I 39, 46
ceramics 60–61
Chest of Kypselos 57
Chrétienne A 39, 46
Cicero 66–67
city xi, 1, 14, 17–18, 56, 62, 84, 86, 91–92, 97, 106
Classical antiquity xii
Classical period xii, xiii, xv, 1–2, 11, 28, 76, 80

Claudius, emperor 59, 68
Clearchus 12
climatology 20
closed harbour 14
coins xv, 62–63
commerce xii, 3, 6
Cyclades, the xvi
Cyprus xii, xv, 5–6, 11–14, 16, 18, 20, 23–24

Darius I 57
data xiv, xv, 1–3, 6, 17, 20–21, 84, 91
decoration xv, xvi, 60, 62, 64, 68
Delphi 57
Delta 94–95, 97, 100, 102, 105–106, 109–110
Demetrios Poliorcetes 28, 63
Demosthenes 12
Diodorus Siculus 12
Dioscorides 12
Dramont A 39

economy xi, xii, xiii, xvi, 1, 4, 75, 80, 105
　economic growth xi, xii
Egypt xiv, xv, 6, 13, 80, 83, 91, 95, 97, 102, 106
Egyptian xiv, 6, 60, 91, 93–94, 96, 98–100, 103
Eighteenth Dynasty xiv, 93
Ephesus 3
Ephorus 12
Ethiopians 13
Euripides xvi, 61, 63
　Iphigenia in Aulis xvi, 61

Fifth Dynasty 102
Flavius Josephus 12
framing xiii, 27, 29–30, 45, 47
freighter xiv, 93

Gaule 84
Gaussian xii
Gela 1 41
Gela 2 43
Geometric period 60–61

Giens peninsula 32
gifts 81
Giglio 40
Goddio, Frank xiv
grain xv, 28, 78, 80–86
Grand Congloué 39
Grand Ribaud F 41
Greco-Roman 27, 32, 40, 46
Greece 83, 92
Greek xiii, xv, 1, 27, 29, 40, 45–46, 56–58, 60, 62, 65, 67, 93, 97, 106, 108

harbour xiv, 3, 13–14, 16–18, 20, 23, 25, 56, 63, 75, 79–80, 84–85, 91, 96–98
　secondary harbours 3
heavy freight xiii
heavy goods 82
Helios 63
Hellenistic period xi, xii, xiii, xiv, 1–6, 11, 13–14, 25, 27–29, 31–32, 35, 39, 43, 45–47, 49, 55–56, 62, 67–68, 75–77, 80–83, 85–86, 108
Heracleion xiv, 6, 91–100, 103–104, 106–109
Heraclides of Macedon 67
Herodotus xiv, xv, 12, 57, 91, 93–95, 102–103, 105, 110
　Historia 91
　Historiae 93
Hesiod 56
Hiero II of Syracuse 28, 64, 81, 85
Hippolytus 17
　Chronicle 17
Homer 13–14, 16, 57, 61
　Iliad 13, 56
　Odyssey 13, 56
hull xiv, xvi, 2–3, 27–28, 31, 39, 43, 45–46, 84, 91, 93, 98–101, 103, 109
Hydra 5

Imperial period xi, xiii, 86
Isis 62
Isocrates 12
Italy 84

itineraries 16–17, 20

Jules-Verne 7 41
Jules-Verne 9 40
Juvenal 66–67

Kargaia 17
Karpaseia 18
Keay, Simon xiv
keel xiii, xiv, 27, 29–30, 36, 43, 46–47, 63, 93–94, 98, 105–106
keelson xiii, 27, 29, 47
Keos xvi
kerkouroi 78, 80, 108
Keryneia 17
Kızılburun 77, 83
Knidos xv, 58
Kouriakos 17
Kourion 23
Kyrenia 2, 31, 39, 43, 45, 77, 108

landmarks 11, 23
Lapathos 17
Late Period 91–92, 97, 106, 109
Latin 66–67
Libya 2
Libyan 13
literary sources xv, 6, 56
Low Aswan Dam 105
luxury goods xiii
Lysippos 59

Ma'agan Mikhael 2, 43
Madrague de Giens 32, 36, 39, 46–47, 49, 83–84
Mahdia 39, 78, 83, 85
Mandrokles of Samos 57
manuscript 11, 17, 25
Marathon 58
marble 68, 83, 86
Marsala 31, 39, 64
Massinissa 81
mast 27, 29, 31, 45, 47, 96, 103–104
Mediterranean xi, xii, xiii, xiv, xvi, 1–6, 11, 13, 27–29, 46–47, 56, 75–78, 80–81, 83, 85–86, 98, 108

ancient Mediterranean xi, 29, 56
Eastern Mediterranean xii, 4–6, 11, 13, 28, 47, 75–77, 80–81, 83, 85
Western Mediterranean 4, 47, 77, 80–81, 83
Melabron 18
Memphis 106, 108–109
Menander 12
merchandise 3, 75, 81
Mersa Gawasis 105
Mikon 58
Misenum 56, 64
mortise and tenon 29

Naples 85
Naucratis 63, 95, 106, 108
Nautical Instructions 13
navigation xiv, xv, 3–4, 6, 11–12, 17–18, 20–21, 56, 91, 94, 96, 98, 100–102, 107, 109
Nealkes of Sicyon 60
New Kingdom 94
Nicias 59
Nicosia xii, 5
Nile xiv, 6, 60, 63, 78, 83, 91–92, 94–98, 100–103, 105, 107–109
    Canopic branch xiv, 91–92, 108
    Nilotic xv, 93, 97, 102
    Upper Nile 98, 103

Old Kingdom 99
Olympia 57–58
Orientalizing ceramics 61
Ostia 49, 107
Ostia Antica 49
Ottoman 23
Ovid 67

Pabuc Burnu 40
painters xv, 55–56, 59–60, 65–68
Paphos 13–14, 17
papyri xiv, 78, 80, 83, 87, 93–94
papyrological 6, 83
Parker, Anthony J. xii, 2
Parrhasios 58
Pausanias 57–58

Persian 60, 108–109
Phaeacians 56
Phaselis 62–63
Philip II 59
Philostratus 59–60
Phoenicia xiii, 13
Phoenician 27, 58
Phoenico-Punic 27, 46
pictorial tradition xv
Piraeus 3
Plane I 39, 46
planking xiii, 27, 29–31, 35, 41, 43, 45, 93–94, 99, 105–106
Plato 12
Pliny xv, 55–56, 59, 64–66
Pointe de Pomègues 39, 46
polis 1
Polybius 12
Polygnotus of Thasos 57
ports xiv, xvi, 13, 16, 78, 95–96, 106–107, 109
Portus Romae xiv
Pozzuoli 3
Propylaea 66
Protogenes of Kaunos xv, 55–56, 65–67
prow xiv, xvi
Ptolemaic 78, 83, 91, 94, 97
Ptolemy I 63
Ptolemy III 85
Ptolemy II Philopator 63
Ptolemy IV Philopator 28
Punic 27–29, 31–32, 46

quadriremes 28
quinqueremes 28

Rhodes 3, 12, 65
Rod el-Air 102
Roman xiii, xiv, 13, 27–29, 32, 40, 46–47, 49, 67–68, 75, 80–84
Roman Republic 27, 29
Rome 59, 67, 76, 80–81, 84, 86
routes xiv, xv, 3, 6, 11–13, 16–17, 20, 23, 75, 83–84

maritime routes 3, 6, 11–13, 16–17, 23
sailors xiii, 13, 16, 20, 25, 57, 62
Salamis 58–59, 63
salinity 105
Samothrace 63, 66
science xii, 57
sea routes xv
Sediment Profile Imaging (SSPI) 96
sewn ships xiii
shell-first xiii
Ship 17 xiv, 6, 93–94, 98–101, 103–106, 109
Ship 43 96, 98
ship archaeologists 5
shipbuilding xii, xiii, xiv, 4, 6, 27–29, 31–32, 39–40, 45, 47, 80, 98
ship painter xv, 55–56, 65, 68
shipworms 105
shipwreck xii, xiii, xiv, xvi, 1–4, 6, 28, 31–32, 36, 39–41, 43, 45–47, 59, 64, 66, 76–78, 80, 83–85, 92
Sinai Peninsula 102
Sixth Dynasty 102
slavery xi, 84
Smyrna 65
Stadiasmus maris Magni xv, 12–14, 16–18, 20–21, 23–24
stern xvi, 29–30, 35–36, 63
stitching xiii
Stoa Poikile 58
Stockholm xvi
stone 75, 81–82, 84–86, 96–97
Strabo xv, 12–14, 16–18, 20–21, 23–24, 84
  Geography 12–14, 16, 21, 23
summer xv, 11, 18, 20
Swedish xvi
Syracuse xvi, 28
Syracusia 28, 31, 36, 84

technology xi, xii, xiii, xiv, 75, 85
  technological change xiv
tenon and mortise xiii, 6
Thalamegos 63

Thasos 57, 79–80, 85
Themetra 49
Theodotus 65
Theopompus 12
Theoros 59
Thonis-Heracleion xiv, 6, 91–93, 95, 103–104, 106–109
Thucydides xvi, 12, 109
Tiber river 107
Timosthenes of Rhodes 12
Titan 39
tonnage xii, xiii, xiv, 6, 28, 34, 46, 75–86, 94, 99, 108–110
trade xii, 1–4, 28, 47, 59, 75, 84–85, 92, 95, 97, 109
  maritime trade xii, 3, 6, 75, 84, 97
  trade relations xii
  trade routes 3
Tretous 23
Triopion xv
triple harbour 14, 17
trireme 28, 46, 109
Trojans 57

Trojan War 57, 59
Tropis conferences 4–5
Tunisia 49
Twenty-seventh Dynasty 109

Vasa xvi
  Vasa Museum xvi
vase xv
Villeneuve-Bargemon 1 41

weather 17–18, 20–21, 101
wheat xv, 28, 75, 81, 83, 85–86
winds xv, 11, 14, 17–18, 20, 23, 25, 65, 94–95, 110
wine xiii, xiv, 29, 31, 43, 46, 75, 81–82, 84, 86
wine-glass profile xiii, 29
winter xv, 11, 18, 20, 108

Xenophon 12

Yassiada 1 28, 47

Zeus 58–59

This book need not end here...

Share

All our books — including the one you have just read — are free to access online so that students, researchers and members of the public who can't afford a printed edition will have access to the same ideas. This title will be accessed online by hundreds of readers each month across the globe: why not share the link so that someone you know is one of them?

This book and additional content is available at:

https://doi.org/10.11647/OBP.0167

Customise

Personalise your copy of this book or design new books using OBP and third-party material. Take chapters or whole books from our published list and make a special edition, a new anthology or an illuminating coursepack. Each customised edition will be produced as a paperback and a downloadable PDF.

Find out more at:

https://www.openbookpublishers.com/section/59/1

Like Open Book Publishers

Follow @OpenBookPublish

Read more at the Open Book Publishers BLOG

You may also be interested in:

Virgil, Aeneid 11 (Pallas & Camilla),
1–224, 498–521, 532–96, 648–89, 725–835
Latin Text, Study Aids with Vocabulary, and Commentary
Ingo Gildenhard and John Henderson

https://doi.org/10.11647/OBP.0158

That Greece Might Still Be Free
The Philhellenes in the War of Independence
William St Clair

https://doi.org/10.11647/OBP.0001

Cicero, Philippic 2, 44–50, 78–92, 100–119
Latin Text, Study Aids with Vocabulary, and Commentary
Ingo Gildenhard

https://doi.org/10.11647/OBP.0156

www.ingramcontent.com/pod-product-compliance
Lightning Source LLC
Chambersburg PA
CBHW041314240426
43669CB00024B/2981